MIMICRY IN BUTTERFLIES

MIMICRY IN BUTTERFLIES

BY

REGINALD CRUNDALL PUNNETT, F.R.S.

Fellow of Gonville and Caius College

Arthur Balfour Professor of Genetics in the University of Cambridge

Cambridge :

at the University Press

1915

CAMBRIDGE
UNIVERSITY PRESS

University Printing House, Cambridge CB2 8BS, United Kingdom

Cambridge University Press is part of the University of Cambridge.

It furthers the University's mission by disseminating knowledge in the pursuit of education, learning and research at the highest international levels of excellence.

www.cambridge.org
Information on this title: www.cambridge.org/9781316601624

© Cambridge University Press 1915

First published 1915
First paperback edition 2015

A catalogue record for this publication is available from the British Library

ISBN 978-1-316-60162-4 Paperback

PREFACE

THIS little book has been written in the hope that it may appeal to several classes of readers.

Not infrequently I have been asked by friends of different callings in life to recommend them some book on mimicry which shall be reasonably short, well illustrated without being very costly, and not too hard to understand. I have always been obliged to tell them that I know of nothing in our language answering to this description, and it is largely as an attempt to remedy this deficiency that the present little volume has been written.

I hope also that it will be found of interest to those who live in or visit tropical lands, and are attracted by the beauty of the butterfly life around them. There are few such countries without some of these cases of close resemblance between butterflies belonging to different families and groups, and it is to those who have the opportunity to be among them that we must look for fuller light upon one of the most fascinating of all nature's problems. If this little book serves to

smooth the path of some who would become acquainted with that problem, and desire to use their
opportunities of observation, the work that has gone to
its making will have been well repaid.

To those who cultivate biological thought from the
more philosophical point of view, I venture to hope
that what I have written may not be without appeal.
At such a time as the present, big with impending
changes in the social fabric, few things are more vital
than a clear conception of the scope and workings
of natural selection. Little enough is our certain
knowledge of these things, and small though the
butterfly's contribution may be I trust that it will
not pass altogether unregarded.

In conclusion I wish to offer my sincere thanks to
those who have helped me in different ways. More
especially are they due to my friends Dr Karl Jordan
for the loan of some valuable specimens, and to
Mr T. H. Riches for his kindly criticism on reading
over the proof-sheets.

R. C. P.

February, 1915

CONTENTS

"The process by which a mimetic analogy is brought about in nature is a problem which involves that of the origin of all species and all adaptations."

H. W. BATES, 1861.

"With mimesis, above all, it is wise, when the law says that a thing is black, first to inquire whether it does not happen to be white."

HENRI FABRE.

CHAPTER I

INTRODUCTORY

IT is now more than fifty years since Darwin gave the theory of natural selection to the world, and the conception of a gradual evolution has long ago become part of the currency of thought. Evolution for Darwin was brought about by more than one factor. He believed in the inherited effects of the use and disuse of parts, and he also regarded sexual selection as operating at any rate among the higher animals. Yet he looked upon the natural selection of small favourable variations as the principal factor in evolutionary change. Since Darwin's time the trend has been to magnify natural selection at the expense of the other two factors. The doctrine of the inherited effects of use and disuse, vigorously challenged by Weismann, failed to make good its case, and it is to-day discredited by the great majority of biologists. Nor perhaps does the hypothesis of sexual selection command the support it originally had. At best it only attempted to explain those features, more especially among the higher animals, in which the sexes differ from one another in pattern, ornament, and the like. With the lapse of time there has come about a tendency to find in natural selection alone a complete explanation of the process of evolution, and to regard it as the sole factor by which all evolutionary change is brought about. Evolution on this view is a gradual process depending upon the slow accumulation by natural selection of small variations, which are more or less inherited, till at last a well-marked change of type is brought about. Could we have before us all the stages through which a given form has passed as natural selection transforms it into another, they would constitute a continuous series such that even refined scrutiny might fail to distinguish between any two consecutive terms. If the slight variations are not of service they will get no favour from natural selection and so can lead to nothing. But if of use in the struggle for existence natural selection preserves them and subsequent variations in the same direction until at length man recognises the accumulation as a new form. Moreover when the perfect thing is once elaborated natural selection will keep it perfect by discouraging any tendency to vary from perfection.

Upon this view, of which the most distinguished protagonist was Weismann, natural selection is the sole arbiter of animal and plant form. Through it and it alone the world has come to be what it is. To it must be ascribed all righteousness, for it alone is the maker. Such in its extreme form is the modern development of Darwin's great contribution to philosophy.

But is it true? Will natural selection really serve to explain all? Must all the various characters of plants and animals be supposed to owe their existence to the gradual operation of this factor working upon small variations ?

Of recent years there has arisen a school of biologists to whom the terms mutationist and Mendelian are frequently applied. Influenced by the writings of Bateson and de Vries, and by the experimental results that have flowed from Mendel's discovery in heredity, they have come to regard the process of evolution as a discontinuous one. The new character that differentiates one variety from another arises suddenly as a sport or mutation, not by the gradual accretion of a vast number of intermediate forms. The white flowered plant has arisen suddenly from the blue, or the dwarf plant from the tall, and intermediates between them need never have existed. The ultimate fate of the new form that has arisen through causes yet unknown may depend upon natural selection. If better endowed than the parent form in the struggle for existence it may through natural selection come to supplant it. If worse endowed natural selection will probably see to its elimination. But if, as may quite possibly happen, it is neither better nor worse adapted than the form from which it sprang, then there would seem to be no reason for natural selection having anything to do with the relation of the new form to its parent.

Between the older and the newer or mutationist point of view an outstanding difference is the rôle ascribed to natural selection. On the one view it

builds up the new variety bit by bit, on the other
the appearance of the new variety is entirely independent of it. From this there follows a radical
difference with regard to the meaning of all the varied
characters of plants and animals. Those who uphold
the all-powerfulness of natural selection are bound
to regard every character exhibited by an animal or
plant as of service to it in the struggle for existence.
Else it could not have arisen through the operation of
natural selection. In other words every character in
plant or animal must be adaptive. On the mutationist
view this of course does not follow. If the new
character which arises independently of natural selection is neither of service nor disservice to its possessors
in the struggle for existence, there seems no reason
why it should not persist in spite of natural selection.
In attempting to decide between the two conflicting
views the study of adaptation is of the first importance.

It was perhaps in connection with adaptation that
Darwin obtained the most striking evidence in support
of his theory, and it is clear from his writings that it
was in this field he laboured with most delight. The
marvellous ways in which creatures may be adapted
in structure and habit for the life they lead had not
escaped the attention of the older naturalists. John
Ray wrote a book[1] upon the subject in which he
pointed out that all things in the Universe, from the
fixed stars to the structure of a bird, or the tongue of

[1] *The Wisdom of God manifested in the Works of the Creation*, London, 1691.

a chameleon, or the means whereby some seeds are wind distributed, are "argumentative of Providence and Design" and must owe their existence to "the Direction of a Superior Cause." Nor have there been wanting other authors who have been equally struck by the wonders of adaptation. But their studies generally led to the same conclusion, an exhortation to praise the infinite Wisdom of Him Who in the days of Creation had taken thought for all these things.

The advent of natural selection threw a new light upon adaptation and the appearance of design in the world. In such books as those on *The Fertilization of Orchids* and *The Forms of Flowers* Darwin sought to shew that many curious and elaborate structures which had long puzzled the botanist were of service to the plant, and might therefore have arisen through the agency of natural selection. Especially was this the case in orchids where Darwin was able to bring forward striking evidence in favour of regarding many a bizarre form of flower as specially adapted for securing the benefits of cross-fertilization through the visits of insects. In these and other books Darwin opened up a new and fascinating field of investigation, and thenceforward the subject of adaptation claimed the attention of many naturalists. For the most part it has been an observational rather than an experimental study. The naturalist is struck by certain peculiarities in the form or colour or habits of a species. His problem is to account for their presence, and as nearly all students of adaptation have been close

followers of Darwin, this generally means an interpretation in terms of natural selection.   Granted this
factor it remains to shew that the character in question
confers some advantage upon the individuals that
possess it.   For unless it has a utilitarian value of some
sort it clearly cannot have arisen through the operation
of natural selection.   However when it comes to the
point direct proof of this sort is generally difficult to
obtain.   Consequently the work of most students of
adaptation consists in a description of the character
or characters studied together with such details of
its life-history as may seem to bear upon the point,
and a suggestion as to how the particular character
studied *may* be of value to its possessors in the struggle
for existence.   In this way a great body of most
curious and interesting facts has been placed on
record, and many ingenious suggestions have been
made as to the possible use of this or that character.
But the majority of workers have taken natural
selection for granted and then interested themselves
in shewing how the characters studied by them might
be of use.   Probably there is no structure or habit
for which it is impossible to devise some use[1], and
the pursuit has doubtless provided many of its devotees
with a pleasurable and often fascinating exercise of
the imagination.   So it has come about that the facts

[1] Ray gives the case of an elephant "that was observed always
when he slept to keep his trunk so close to the ground, that nothing
but Air could get in between them," and explains it as an adaptation
in habit to prevent the mice from crawling into its lungs—"a strange
sagacity and Providence in this Animal, or else an admirable instinct."

instead of being used as a test of the credibility of natural selection, serve merely to emphasise the pæan of praise with which such exercises usually conclude. The whole matter is too often approached in much the same spirit as that in which John Ray approached it two centuries ago, except that the Omnipotency of the Deity is replaced by the Omnipotency of Natural Selection. The vital point, which is whether Natural Selection *does* offer a satisfactory explanation of the living world, is too frequently lost sight of. Whether we are bound or not to interpret all the phenomena of life in terms of natural selection touches the basis of modern philosophy. It is for the biologist to attempt to find an answer, and there are few more profitable lines of attack than a critical examination of the facts of adaptation. Though "mimicry" is but a small corner in this vast field of inquiry it is a peculiarly favourable one owing to the great interest which it has excited for many years and the consequently considerable store of facts that has been accumulated. If then we would attempt to settle this most weighty point in philosophy there is probably nothing to which we can appeal with more confidence than to the butterfly.

CHAPTER II

MIMICRY is a special branch of the study of adaptation. The term has sometimes been used loosely to include cases where an animal, most frequently an insect, bears a strong and often most remarkable resemblance to some feature of its inanimate surroundings. Many butterflies with wings closed are wonderfully like dead leaves ; certain spiders when at rest on a leaf look exactly like bird-droppings ; "looper" caterpillars simulate small twigs ; the names of the "stick-" and "leaf-" insects are in themselves an indication of their appearance. Such cases as these, in which the creature exhibits a resemblance to some part of its natural surroundings, should be classified as cases of "protective resemblance" in contradistinction to mimicry proper. Striking examples of protective resemblance are abundant, and though we possess little critical knowledge of the acuity of perception in birds and other insect feeders it is plausible to regard the resemblances as being of definite advantage in the struggle for existence. However, it is with mimicry and not with protective coloration in general that we are here directly

concerned, and the nature of the phenomenon may
perhaps best be made clear by a brief account of the
facts which led to the statement of the theory.

In the middle of last century the distinguished
naturalist, H. W. Bates, was engaged in making
collections in parts of the Amazon region. He paid
much attention to butterflies, in which group he
discovered a remarkably interesting phenomenon[1].
Among the species which he took were a large number
belonging to the group Ithomiinae, small butterflies
of peculiar appearance with long slender bodies and
narrow wings bearing in most cases a conspicuous
pattern (cf. Pl. X, fig. 7). When Bates came to
examine his catch more closely he discovered that
among the many Ithomiines were a few specimens
very like them in general shape, colour, and markings,
but differing in certain anatomical features by
which the Pierinae, or "whites," are separated from
other groups. Most Pierines are very different from
Ithomiines. It is the group to which our common
cabbage butterfly belongs and the ground colour is
generally white. The shape of the body and also of the
wings is in general quite distinct from what it is in the
Ithomiines. Nevertheless in these particular districts
certain of the species of Pierines had departed widely
from what is usually regarded as their ancestral
pattern (Pl. X, fig. 1) and had come to resemble very
closely the far more abundant Ithomiines among whom
they habitually flew (cf. Pl. X, figs. 2 and 3). To

[1] *Trans. Linn. Soc.* vol. 23, 1862.

use Bates' term they "mimicked" the Ithomiines, and he set to work to devise an explanation of how this could have come about. The *Origin of Species* had just appeared and it was natural that Bates should seek to interpret this peculiar phenomenon on the lines there laid down. How was it that these Pierines had come to depart so widely from the general form of the great bulk of their relations, and to mimic so closely in appearance species belonging to an entirely different group, while at the same time conserving the more deeply seated anatomical features of their own family ? If the change was to be regarded as having come about through the agency of natural selection it must clearly be of advantage to the mimicking forms ; otherwise natural selection could not come into operation. What advantage then have the Ithomiines over the majority of butterflies in those parts ? They are small insects, rather flimsy in build, with comparatively weak powers of flight, and yet so conspicuously coloured that they can hardly be mistaken for anything else. In spite of all this they are little subject to the attacks of enemies such as birds, and Bates attributed this to the fact that the juices of their bodies are unpalatable. According to him their striking and conspicuous pattern is of the nature of a warning coloration, advertising their disagreeable properties to possible enemies. A bird which had once attempted to eat one would find it little to its taste. It would thenceforward associate the conspicuous pattern with a disagreeable flavour and in future leave such butterflies severely alone. The more conspicuous the pattern the more readily would it be noticed by the enemy, and so it would be of advantage to the Ithomiine to possess as striking a pattern as possible. Those butterflies shewing a tendency to a more conspicuous pattern would be more immune to the attacks of birds and so would have a better chance of leaving progeny than those with a less conspicuous pattern. In this way variations in the direction of greater conspicuousness would be accumulated gradually by natural selection, and so would be built up in the Ithomiine the striking warning coloration by which it advertises its disagreeable properties. Such is the first step in the making of a mimicry case—the building up through natural selection of a conspicuous pattern in an unpalatable species by means of which it is enabled to advertise its disagreeable properties effectively and thereby secure immunity from the attacks of enemies which are able to appreciate the advertisement. Such patterns and colours are said to be of a "warning" nature. The existence of an unpalatable model in considerable numbers is the first step in the production of a mimetic resemblance through the agency of natural selection.

We come back now to our Pierine which must be assumed to shew the general characters and coloration of the family of whites to which they belong (cf. Pl. X, fig. 1). Theoretically they are not specially protected by nauseous properties from enemies and hence their conspicuous white coloration renders them especially liable to attack. If, however, they could exchange their normal dress for one resembling that of the Ithomiines it is clear that they would have a chance of being mistaken for the latter and consequently of being left alone. Moreover, in certain cases these Pierines *have* managed to discard their normal dress and assume that of the Ithomiines. On theoretical grounds this must clearly be of advantage to them, and being so might conceivably have arisen through the operation of natural selection. This indeed is what is supposed to have taken place on the theory of mimicry. Those Pierines which exhibited a variation of colour in the direction of the Ithomiine "model" excited distrust in the minds of would-be devourers, who had learned from experience to associate that particular type of coloration with a disagreeable taste. Such Pierines would therefore have a rather better chance of surviving and of leaving offspring. Some of the offspring would exhibit the variation in a more marked degree and these again would in consequence have a yet better chance of surviving. Natural selection would encourage those varying in the direction of the Ithomiine model at the expense of the rest and by its continuous operation there would gradually be built up those beautiful cases of resemblance which have excited the admiration of naturalists.

Wallace was the next after Bates to interest himself in mimicry and, from his study of the butterflies of the Oriental region[1], shewed that in this part of

[1] *Trans. Linn. Soc.* vol. 25, 1866.

the world too there existed these remarkable resemblances between species belonging to different families. Perhaps the most important part of Wallace's contribution was the demonstration that in some species not only was it the female alone that "mimicked" but that there might be several different forms of female mimicking different models, and in some cases all unlike the male of their own species. One of the species studied by Wallace, *Papilio polytes*, is shewn on Plate V. We shall have occasion to refer to this case later on, and it is sufficient here to call attention to the three different forms of female, of which one is like the male while the other two resemble two other species of *Papilio*, *P. hector* and *P. aristolochiae*, which occur in the same localities. Instances where the female alone of some unprotected species mimics a model with obnoxious properties are common in all tropical countries. It has been suggested that this state of things has come about owing to the greater need of protection on the part of the female. Hampered by the disposal of the next generation the less protected female would be at a greater disadvantage as compared with the mimic than would the corresponding male whose obligations to posterity are more rapidly discharged. The view of course makes the assumption that the female transmits her peculiar properties to her daughters but not to her sons.

A few years later Trimen[1] did for Africa what Bates had done for America and Wallace for Indo-

---

[1] *Trans. Linn. Soc.* vol. 26, 1870.

Malaya. It was in this paper that he elucidated that most remarkable of all cases of mimicry—*Papilio dardanus* with his harem of different consorts, all tailless, all unlike himself, and often wonderfully similar to unpalatable forms found in the same localities (cf. p. 30).

We may now turn to one of the most ingenious developments of the theory of mimicry. Not long after Bates' original memoir appeared attention was directed to a group of cases which could not be explained on the simple hypothesis there put forward. Many striking cases of resemblance had been adduced in which both species obviously belonged to the presumably unpalatable groups. Instances of the sort had been recorded by Bates himself and are perhaps most plentiful in South America between species belonging respectively to the Ithomiinae and Heliconinae. On the theory of mimicry all the members of both of these groups must be regarded as specially protected owing to their conspicuous coloration and distasteful properties. What advantage then can an Ithomiine be supposed to gain by mimicking a Heliconine, or *vice versâ*? Why should a species exchange its own bright and conspicuous warning pattern for one which is neither brighter nor more conspicuous? To Fritz Müller, the well-known correspondent of Darwin, belongs the credit of having suggested a way out of the difficulty. Müller's explanation turns upon the education of birds. Every year there hatch into the world fresh generations of young birds, and each generation has to learn afresh from experience what is pleasant to eat and what is not. They will try all things and hold fast to that which is good. They will learn to associate the gay colours of the Heliconine and the Ithomiine with an evil taste[1] and they will thenceforward avoid butterflies which advertise themselves by means of these particular colour combinations. But in a locality where there are many models, each with a different pattern and colour complex, each will have to be tested separately before the unpalatableness of each is realised. If for example a thousand young birds started their education on a population of butterflies in which there were five disagreeable species, each with a distinct warning pattern, it is clear that one thousand of each would devote their lives to the education of these birds, or five thousand butterflies in all[2]. But if these five species, instead of shewing five distinct warning patterns, all displayed the same one it is evident that the education of the birds would be accomplished at the price of but one thousand butterfly existences instead of five. Even if one of the five species were far more abundant than the others it would yet be to its advantage that the other four should exhibit the same warning pattern. Even though the losses were distributed *pro rata* the more abundant species would profit to some extent. For

[1] In attributing this quality to the butterflies in question I am merely stating what is held by the supporters of the mimicry theory. I know of scarcely any evidence either for or against the supposition.

[2] It is assumed that the intelligence of the birds is such that they can learn a pattern after a single disagreeable experience of it.

the less abundant species the gain would of course be relatively greater. Theoretically therefore, all of the five species would profit if in place of five distinct warning patterns they exhibited but a single one in common. And since it is profitable to all concerned what more natural than that it should be brought about by natural selection ?

Müller's views are now widely accepted by students of mimicry as an explanation of these curious cases where two or more evidently distasteful species closely resemble one another. Indeed the tendency in recent years has been to see Müllerian mimicry everywhere, and many of the instances which were long regarded as simple Batesian cases have now been relegated to this category. The hypothesis is, of course, based upon what appears to man to be the natural behaviour of young birds under certain conditions. No one knows whether young birds actually do behave in the way that they are supposed to. In the absence of any such body of facts the Müllerian hypothesis cannot rank as more than a plausible suggestion, and, as will appear later, it is open to severe criticism on general grounds.

Perhaps the next contribution to the subject of mimicry which must rank of the first importance was that of Erich Haase[1], to whose book students of these matters must always be under a heavy obligation. It was the first and still remains the chief work of general scope. Since Haase's day great numbers of

---

[1] *Untersuchungen über die Mimikry*, 1893.

fresh instances of mimetic resemblance have been recorded from all the great tropical areas of the world, and the list is being added to continually. Most active in this direction is the Oxford School under Professor Poulton to whose untiring efforts are largely due the substantial increases in our knowledge of African butterflies contributed by various workers in the field during the past few years. Whatever the interpretation put upon them, there can be no question as to the value of the facts brought together, more especially those referring to the nature of the families raised in captivity from various mimetic forms. With the considerable additions from Africa[1] during the past few years several hundreds of cases of mimicry must now have been recorded. Some of the best known and most striking from among these will be described briefly in the next two chapters.

---

[1] The African mimetic butterflies have been recently monographed by Eltringham in a large and beautifully illustrated work—*African Mimetic Butterflies*, Oxford, 1910.

CHAPTER III

OLD-WORLD MIMICS

THE earlier naturalists who studied butterflies made use of colour and pattern very largely in arranging and classifying their specimens. Insects shewing the same features in these respects were generally placed together without further question, especially if they were known to come from the same locality. In looking through old collections of butterflies from the tropics it is not infrequent to find that the collector was deceived by a mimetic likeness into placing model and mimic together. During the last century, however, more attention was paid to the anatomy of butterflies, with the result that their classification was placed upon a basis of structure. As in all work of the sort certain features are selected, partly owing to their constancy and partly for their convenience, the insects being arranged according as to whether they present these features or not. Everybody knows that the butterflies as a group are separated from the moths on the ground that their antennae are club shaped at the end, while those of the moth are generally filamentary and taper to a fine point. The butterflies themselves

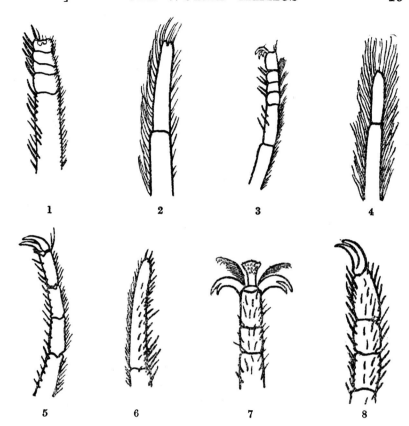

Figs. 1—8. Terminal portion of front legs of butterflies belonging
to different families. (After Eltringham.)

1. *Hypolimnas misippus,* ♀ (Nymphalidae).
2.   ,,   ,,   ♂ (,,   ).
3. *Abisara savitri,*   ♀ (Erycinidae).
4.   ,,   ,,   ♂ (,,   ).
5. *Lycaena icarus,*   ♀ (Lycaenidae).
6. *Cupido zoë,*   ♂ (,,   ).
7. *Ganoris rapae,*   ♂ (Pieridae).
8. *Papilio echerioides,* ♀ (Papilionidae).

may be subdivided into five main groups or families[1] according to the structure of the first of their three pairs of legs. In the Papilionidae or "swallow-tails," the first pair of legs is well developed in both sexes (Fig. 8). In the Pieridae or "whites," the front legs are also similar in both sexes, but the claws are bifid and a median process, the empodium, is found between them (Fig. 7). In the remaining three families the front legs differ in the two sexes. The females of the Lycaenidae or "blues" have well-developed front legs in which the tarsus is terminated by definite claws (Fig. 5), whereas in the males the terminal part of the leg, or tarsus, is unjointed and furnished with but a single small claw (Fig. 6). This reduction of the front legs has gone somewhat further in the Erycinidae (Figs. 3 and 4), a family consisting for the most part of rather small butterflies and specially characteristic of South America. In the great family of the Nymphalidae the reduction of the front legs is well marked in both sexes. Not only are they much smaller than in the other groups, but claws are lacking in the female as well as in the male (Figs. 1 and 2).

Though the structure of the fore limbs is the character specially chosen for separating these different families from one another, it is of course understood that they differ from one another in various other distinctive features. The chrysalis of the Nymphalidae for example hangs head downwards suspended by the

---

[1] Omitting the Hesperidae which hardly enter into questions of mimicry.

tail, whereas in the Pieridae and Papilionidae meta-
morphosis takes place with the chrysalis attached by
the tail but supported also by a fine girdle of silk
round the middle so that the head is uppermost. The
larvae also afford characters by which some of the
families may be distinguished—those of the Papilionidae
for example having a process on the back which can
be extruded or retracted.

Owing to the great size of the family of the Nymphalidae, in which the number of species approaches
5000, it is convenient to deal with the eight sub-groups
into which it has been divided. The characters serving
to mark off the sub-groups from one another are various.
Sometimes it is the minuter structure of the tarsus, at
others the form of the caterpillar or the chrysalis, at
others the arrangement of the nervures that form the
skeleton of the wing. Into these systematic details,
however, we need not enter more fully here[1]. What is
important from the standpoint of mimicry is that
these divisions, made solely on anatomical structure,
correspond closely with the separation of models from
mimics. Of the eight sub-families into which the
Nymphalidae are divided four, viz. the Danainae,
Acraeinae, Heliconinae, and Ithomiinae, provide models
and some, but far fewer, mimics ; two, the Satyrinae
and Nymphalinae, provide many mimics and but few
models, while two groups, the Morphinae and Bras-
solinae, practically do not enter into the mimicry story.

---

[1] The classification adopted is that used by Dr Sharp in the
"Cambridge Natural History," *Insects*, vol. 2, 1901.

Simple mimicry, explicable, at any rate in theory, on the lines laid down by Bates, is a phenomenon of not infrequent occurrence in tropical countries, though rare in more temperate lands. In each of the three great divisions of the tropical world we find certain groups of butterflies serving as models, and being mimicked by butterflies belonging as a rule to quite different groups. Speaking generally the models of any given region are confined to a few groups, while the mimics are drawn from a greater number. In Asia the principal models belong to the Danaines, the Euploeines, and to a group of swallow-tails which from the fact that their larvae feed on the poisonous *Aristolochia* plant are generally distinguished as the "Poison-eaters," or *Pharmacophagus* group. Of these the Danaines and Euploeines are closely related and have much in common. They are usually butterflies of medium size, of rather flimsy build and with a somewhat slow and flaunting flight. In spite, however, of their slight build they are toughly made and very tenacious of life. Most butterflies are easily killed by simply nipping the thorax. There is a slight crack and the fly never recovers. But the collector who treats a Danaid in a way that would easily kill most butterflies is as likely as not many hours after to find it still alive in his collecting box or in the paper to which it may have been transferred when caught. They give one the impression of being tougher and more "rubbery" in consistence than the majority of Lepidoptera. Moreover, the juices of their bodies seem to be more oily and less easily dried up. In general colour scheme they vary a great deal. Some, such as *Danais chrysippus* (Pl. IV, fig. 1), are conspicuous with their bright fulvous-brown ground colour and the sharp white markings on the black tips of their fore wings. Others again such as *Danais septentrionis* (Pl. I, fig. 3), with a dark network of lines on a pale greenish ground, are not nearly so conspicuous. Of the Euploeines some have a beautiful deep blue metallic lustre (cf. Pl. II, fig. 4), though many are of a plain sombre brown relieved only by an inconspicuous border of lighter markings (cf. Pl. I, fig. 10).

Both Danaines and Euploeines serve as models for a great variety of species belonging to different groups. *Danais septentrionis* (Pl. I, fig. 3) is a very abundant species in India and Ceylon, and in the same region there are several other very similar species. Flying with them in Northern India are two species of *Papilio*, *P. macareus* and *P. xenocles* (Pl. I, fig. 4), which resemble these Danaids fairly closely. In Southern India and Ceylon one of the two forms of *Papilio clytia* (Pl. I, fig. 7) is also regarded as a mimic of these Danaids. In the same part of the world there is a Pierine of the genus *Pareronia*, whose female is very like these Danaines on the upper surface (Pl. I, fig. 1). The male of this Pierine is quite distinct from the female (Pl. I, fig. 2).

The common *Danais chrysippus* (Pl. IV, fig. 1), found in this region, has been described as probably the most abundant butterfly in the world, and serves

as a model for several species belonging to different
groups. It and its mimics will, however, be described
in more detail later on. Mention must also be made
of the striking case of the Danaid, *Caduga tytia* and
its Papilionine mimic *P. agestor* from Sikkim (Pl. II,
figs. 2 and 3). In both species the fore wings are
pale blue broken by black; while the hind wings are
pale with a deep outer border of rusty red. Not only
in colour but also in shape the swallow-tail bears a
remarkable resemblance to the Danaid. *C. tytia* is
also mimicked by a rare Nymphaline *Neptis imitans*,
which exhibits the same striking colour scheme so very
different from that of most of its allies.

No less remarkable are some of the cases in which
the Euploeines serve as models. *E. rhadamanthus*, for
example, is mimicked by the scarce *Papilio mendax*,
and a glance at Figs. 8 and 9 on Plate II shews
how well this butterfly deserves its name. *Euploea
rhadamanthus* also serves as a model for one of the
several forms of female of the Nymphaline species
Euripus halitherses. In some Euploeines the sexes are
different in appearance—a somewhat unusual thing
among butterflies serving as models in cases of mimetic
resemblance. Such a difference is found in *Euploea
mulciber*, the male being predominantly brown with a
beautiful deep blue suffusion, while the female is a
rather lighter insect with less of the blue suffusion
and with hind wings streaked with lighter markings
(Pl. II, figs. 4 and 5). It is interesting to find that
Elymnias malelas, a Satyrid which mimics this species,

shews a similar difference in the two sexes (Pl. II, figs. 6 and 7).

It is remarkable that similar sexual difference is also shewn by the rare *Papilio paradoxus*, the two sexes here again mimicking respectively the two sexes of *Euploea mulciber*.

Many of the Euploeines, more especially those from Southern India and Ceylon, lack the blue suffusion, and are sombre brown insects somewhat relieved by lighter markings along the hinder border of the hind wings. *Euploea core* (Pl. I, fig. 10), a very common insect, is typical of this group. A similar coloration is found in one of the forms of *Papilio clytia* (Pl. I, fig. 8) from the same region as well as in the female of the Nymphaline species *Hypolimnas bolina* (Pl. I, fig. 6). The male of this last species (Pl. I, fig. 5) is quite unlike its female, but is not unlike the male of the allied species, *H. misippus*, which it resembles in the very dark wings each with a white patch in the centre, the junction of light and dark being in each case marked by a beautiful purple-blue suffusion. There is also a species of *Elymnias* (*E. singhala*) in this part of the world which in general colour scheme is not widely dissimilar from these brown Euploeas (Pl. I, fig. 9).

The third main group of models characteristic of this region belongs to the Papilionidae. It was pointed out by Haase some 20 years ago that this great family falls into three definite sections, separable on anatomical grounds (see Appendix II). One of these sections he termed the *Pharmacophagus* or "poison-eating" group owing to the fact that the larvae feed on the poisonous climbing plants of the genus *Aristolochia*. It is from this group that all Papilios which serve as models are drawn. No mimics of other unpalatable groups such as Danaines are to be found among the Oriental Poison-eaters. In the other two sections of the genus mimics are not infrequent (cf. Appendix II), though probably none of them serve as models. To the Pharmacophagus group belong the most gorgeous insects of Indo-Malaya—the magnificent Ornithoptera, largest and most splendid of butterflies. It is not a large proportion of the members of the group which serve as models, and these on the whole are among the smaller and less conspicuous forms. In all cases the mimic, when a butterfly, belongs to the *Papilio* section of the three sections into which Haase divided the family (cf. Appendix II). *Papilio aristolochiae* (Pl. V, fig. 5), for example, is mimicked by a female form of *Papilio polytes,* and the geographical varieties of this widely spread model are generally closely paralleled by those of the equally wide spread mimic. For both forms range from Western India across to Eastern China. Another poison-eater, *P. coon,* provides a model for one of the females of the common *P. memnon.* It is curious that in those species of the poison-eaters which serve as models the sexes are practically identical in pattern, and are mimicked by certain females only of the other two Papilio groups, whereas in the Ornithoptera, which also belong to the poison-eaters, the difference between the sexes is exceedingly striking.

Though the Pharmacophagus Papilios are mimicked only by other Papilios among butterflies they may serve occasionally as models for certain of the larger day-flying moths. *Papilio polyxenus*, for example, is mimicked not only by the unprotected *P. bootes* but also by the moth *Epicopeia polydora* (Pl. III, figs. 5 and 6). Like the butterfly the *Epicopeia*, which is comparatively rare, has the white patch and the outer border of red marginal spots on the hind wing. Though it is apparently unable to provide itself with an orthodox tail it nevertheless makes a creditable attempt at one. There are several other cases of mimetic resemblance between day-flying moths and Pharmacophagus swallow-tails—the latter in each case serving as the model. Rarely it may happen that the rôle of butterfly and moth is reversed, and the butterfly becomes the mimic. A very remarkable instance of this is found in New Guinea where the rare *Papilio laglaizei* mimics the common day-flying moth *Alcidis agathyrsus*. Viewed from above the resemblance is sufficiently striking (Pl. III, figs. 1 and 2), but the most wonderful feature concerns the underneath. The ventral half of the moth's abdomen is coloured brilliant orange. When the wings are folded back they cover and hide from sight only the dorsal part of the abdomen, so that in this position the orange neutral surface is conspicuous. When, however, the wings of the butterfly are folded they conceal the whole of the abdomen. But the butterfly has developed on each hind wing itself a bright orange patch in such a position that when the wings are folded back the orange patch lies over the sides of the abdomen. In this way is simulated the brilliant abdomen of the moth by a butterfly, in which, as in its relations, this part is of a dark and sombre hue.

A few models are also provided in the Oriental region by the genus *Delias*, which belongs to the Pierines. A common form, *Delias eucharis*, is white above but the under surface of the hind wings is conspicuous with yellow and scarlet (Pl. II, fig. 1). It has been suggested that this species serves as a model for another and closely allied Pierine, *Prioneris sita*, a species distinctly scarcer than the *Delias*. There is some evidence that the latter is distasteful (cf. p. 115), but nothing is known of the *Prioneris* in this respect. Other species of *Delias* are said to function as models for certain day-flying moths belonging to the family Chalcosiidae, which may bear a close resemblance to them. In certain cases it may happen that the moth is more abundant than the Pierine that it resembles[1].

Tropical Africa is probably more wealthy in mimetic analogies than Indo-Malaya, and the African cases have recently been gathered together by Eltringham in a large and beautifully illustrated memoir[2]. The principal models of the region are furnished by the Danainae and the allied group of the Acraeinae. Of the Danaines one well-known model, *Danais chrysippus*,

[1] Cf. Shelford, *Proc. Zool. Soc.* 1902.
[2] *African Mimetic Butterflies*, Oxford, 1910.

is common to Africa and to Indo-Malaya. Common also to the two regions are the mimics, *Argynnis hyperbius* and *Hypolimnas misippus* (cf. Pl. IV, figs. 3 and 7). The case of the last named is peculiarly interesting because it presents well-marked varieties which can be paralleled by similar ones in *D. chrysippus*. In addition to the typical form with the dark tipped fore wing relieved by a white bar there is in each species a form uniformly brown, lacking both the dark tip and the white bar of the fore wing. There is also another form in the two species in which the hind wing is almost white instead of the usual brown shade. In both species, moreover, the white hind wing may be associated either with the uniformly brown fore wing or with the typical form. There is also another common African butterfly, *Acraea encedon*, in which these different patterns are closely paralleled (cf. Pl. IX). Several other species of butterflies and a few diurnal moths bear a more or less close resemblance to *D. chrysippus*.

Danaine butterflies with the dark interlacing lines on a pale greenish-blue ground, so characteristic of the Oriental region, are represented in Africa by the species *Danais petiverana* (Pl. VI, fig. 1) ranging across the continent from Sierra Leone to British East Africa. A common Papilio, *P. leonidas* (Pl. VI, fig. 2) has a similar extensive range, and has been regarded as a mimic of the Danaine. In S. Africa *P. leonidas* is represented by the variety *brasidas* in which the white spots are reduced and the blue-green ground is lacking. *Brasidas* bears a strong resemblance to the tropical

Danaine *Amauris hyalites* (Pl. VI, fig. 3) of which it has been regarded as a mimic. It must however be added that it is only over a small part of their respective ranges, viz. in Angola, that the two species are to be met with together.

The butterflies belonging to the genus *Amauris* are among the most abundant and characteristic Danaine models of Africa. Some of the black and white species such as *A. niavius* (Pl. VIII, fig. 6) are conspicuous insects in a cabinet. Others again, such as *A. echeria* (Pl. VIII, fig. 7), are relatively sombre-looking forms. Among the best known mimics of the genus is a species of *Hypolimnas*[1]—*H. dubius*. This interesting form is polymorphic and mimics different species of *Amauris*. The variety *wahlbergi*, for example, is very like *A. niavius*, while *mima* strongly resembles *A. echeria* (Pl. VIII, figs. 8 and 9). It was at one time supposed that these two varieties of *Hypolimnas dubius* were different species and the matter was only definitely settled when the two forms were bred from the eggs of the same female. Other mimics of *Amauris* are found among the Papilios and the Nymphaline genus *Pseudacraea*.

But among all the mimics of Danaines in Africa and elsewhere *Papilio dardanus* is pre-eminent, and has been described by more than one writer as the most important case of mimicry in existence. Not only does it shew remarkable resemblances to various

[1] These African species of *Hypolimnas* are frequently referred to the genus *Euralia*.

Danaids, but it presents features of such peculiar interest that it must be considered in more detail. *Papilio dardanus* in its various sub-races is spread over nearly all the African continent south of the Sahara. Over all this area the male, save for relatively small differences, remains unchanged—a lemon-yellow insect, tailed, and with black markings on fore and hind wings (Pl. VIII, fig. 1). The female, however, exhibits an extraordinary range of variation. In South Africa she appears in three guises, (1) the *cenea* form resembling *Amauris echeria*, (2) the *hippocoon* form like *Amauris niavius*, and (3) the *trophonius* form which is a close mimic of the common *Danais chrysippus*[1]. Except that *cenea* does not occur on the West Coast these three forms of female are found over almost all the great continental range of *dardanus* and its geographical races. Northwards in the latitude of Victoria Nyanza occurs a distinct form of female, *planemoides*, which bears a remarkable resemblance to the common and distasteful *Planema poggei*, and is found only where the latter is abundant. All of these four forms are close mimics of a common Danaine or Acraeine model. Other forms of female, however, are known, of which two, *dionysus* and *trimeni*, are sufficiently distinct and constant to have acquired special names. *Dionysus* may be said to unite the fore wing of the *hippocoon* form with the hind wing of the *trophonius* form, except that the colour of the last part is yellow instead of

---

[1] Corresponding to the *dorippus* form of *D. chrysippus* (cf. Pl. IX) there is a rare form of *trophonius* known as *dorippoides*.

bright brown. It is a western form and is unlike any model. *Trimeni* also is unlike any model but is of peculiar interest in that it is much more like the male with its pale creamy-yellow colour and the lesser development of black scales than occurs in most of the forms of female. At the same time the general arrangement of the darker markings is on the whole similar to that in the *hippocoon* and in the *trophonius* form. *Trimeni* is found on the Kikuyu Escarpment, near Mt Kenia, along with the four mimicking forms.

Continental Africa, south of the equator, has produced no female similar to the male. But in Abyssinia is found another state of things. Here, so far as is known, occur three forms, all tailed, of which one is similar in general colour and pattern to the male, while the other two, *niavioides* and *ruspina*[1], resemble respectively a tailed *hippocoon* and a tailed *trophonius*. Lastly we have to record that *Papilio dardanus* is also found as the geographical race *humbloti* on Comoro Island, and as *meriones* on Madagascar. In both forms the females are tailed, and resemble the males.

From this long series of facts it is concluded that the male of *P. dardanus* represents the original form of both sexes. On the islands of Comoro and Madagascar this state of things still survives. But it is supposed that on the African continent existed enemies which persecuted the species more than on the islands

[1] These two forms are figured on Plate 10 of Eltringham's *African Mimetic Butterflies*.

and encouraged the development of mimetic forms in the female. The original female still lingers in Abyssinia though it is now accompanied by the two mimetic forms *niavioides* and *ruspina*. Over the rest of the area occupied by *dardanus* the females are always tailless and, with the exception of *trimeni* and *dionysus*, wonderfully close mimics. *Trimeni*, the intermediate form, provides the clue to the way in which the mimetic females have been derived from the male, viz. by the prolongation across the fore wing of the dark costal bar already found in the females of the Madagascar and Abyssinian races, by the deepening of the dark edging to the wings, and by the loss of the tail. Through the gradual accumulation of small variations *trimeni* came from the male-like female, and by further gradual accumulation of small favourable variations the mimetic forms came from *trimeni*. South of the equator the male-like form and the intermediate *trimeni* have disappeared owing to the stringency of selection being greater. Moreover the likeness of mimic to model is closer than in the north, a further proof of the greater stringency of natural selection in these parts. Such in brief is the explanation in terms of mimicry of the remarkable and complex case of *dardanus*.

Although the Euploeinae are not represented on the African continent, it is the headquarters of another distasteful family of butterflies—the Acraeinae—which is but sparingly represented in the Oriental region[1].

[1] *Acraea violae*, the only representative of the group in S. India

Of smaller size than the Danaines they are characterised, like this group, by their tenacity of life and
by the presumably distasteful character of their
body juices. They are said also to possess an offensive
odour apparently exuded through the thorax. The
majority of the members of the group fall into the
two genera *Acraea* and *Planema*. Species of Acraea
are on the whole characterised by their general bright
red-brown colour and by the conspicuous black spots
on both fore and hind wings. A typical Acraeine
pattern is that of *Acraea egina* (Pl. VI, fig. 7) which
is mimicked remarkably closely by the Nymphaline
Pseudacraea boisduvali and by the Swallow-tail *Papilio
ridleyanus* (Pl. VI, figs. 5 and 6).

In the genus *Planema* the spots are as a rule fewer
and clustered near the body, while on both fore and
hind wings there is a tendency to develop clear wide
band-like areas of orange or white (cf. Pl. VII).

Like the Acraeas the Planemas are principally
mimicked by species of *Pseudacraea* and of *Papilio*.
Some of the cases of resemblance between *Planema*
and *Pseudacraea* are among the most striking known.
Planema macarista is one of those comparatively rare
instances in which a model shews a marked difference
in the pattern of the two sexes. The clear area on the
fore wing of the male is deep orange, whereas in the
female it is somewhat different in shape, and, like the
area on the hind wing, is white (cf. Pl. VII, figs. 1 and 2).

and Ceylon, is nevertheless a very abundant insect. It cannot, however,
be said that it is definitely mimicked by any other species in this region.

Pseudacraea eurytus hobleyi (Pl. VII, figs. 6 and 7) shews a similar difference in the sexes, the male and female of this species mimicking respectively the male and female of *Planema macarista*. The case is made even more remarkable by the fact that both of the sexual forms of *Planema macarista* are mimicked by the Satyrine *Elymnias phegea* (Pl. VII, fig. 9), though in this species either the black and white, or the black, white, and orange form may occur in either sex. Among the best Papilionine mimics of the Planemas is *Papilio cynorta* whose female is extraordinarily like the common *Planema epaea* (Pl. VII, figs. 5 and 10). The resemblance of the *planemoides* female of *P. dardanus* to *P. poggei* has already been noticed.

A striking feature of the African continent is the frequency with which mimetic forms are found among the Lycaenidae. As a rule the "blues" rarely exhibit mimetic analogies, but in Africa there are several species, especially those of the genus *Mimacraea*, which closely resemble Acraeines. Others again bear a marked resemblance to certain small Pierines, *Citronophila similis* from S. Nigeria for example being extraordinarily like the common *Terias brigitta*, a small bright yellow Pierine with black-edged wings.

A remarkable feature of the African continent is the absence of the Pharmacophagus Swallow-tails. Of such Papilios as exhibit mimicry, and as compared with the total number of the group present the proportion is large, the majority resemble one or other of the characteristic Danaines, while a few such as *P. ridleyanus* and *P. cynorta* resemble either an Acraeoid or a Planemoid model.

As in the Oriental region the African Pierines do not offer many instances of mimetic analogies. The genus *Mylothris*, in which certain species are characterised by orange patches at the bases of the undersurfaces of the fore wings, is regarded by some authors as providing models for allied genera such as *Belenois* and *Phrissura*. But as neither models nor mimics offer a marked divergence in appearance from the ordinary Pierine facies it is doubtful whether much stress can be laid on these cases.

Africa also offers a few striking instances of mimicry in which day-flying moths play a part. The conspicuous Geometer *Aletis helcita* is an abundant form, and with its strong red colour and black wing margins broken by white it is a striking object in the preserved state. Among the forms which bear a close resemblance to it are the Nymphaline *Euphaedra ruspina*, and the Lycaenid *Telipna sanguinea*[1].

[1] Coloured figures of these and of the other African species referred to may be found in Eltringham's work on *African Mimetic Butterflies*.

CHAPTER IV

NEW-WORLD MIMICS

OF all the continents South America affords the greatest wealth of butterfly life, and it is in the tropical part of this region that many of the most beautiful and striking cases of mimicry are to be found. Viewed as a whole the butterfly population presents several features which serve to mark it off from that of the other two great tropical areas. In the first place the proportion of gaily coloured forms is higher. Bright red, yellow or fulvous brown contrasted with some deep shade approaching black form the dominant notes. Sombre coloured species are relatively scarcer than in the Oriental and African regions. In the second place when looking over collections from this part of the world one cannot help being struck by the frequency with which similar colour combinations occur over and over again in different as well as in the same groups. Now it is a simple scheme of black with an oblique scarlet band upon the fore wings—now an arrangement with alternating stripes of bright brown and black relieved with patches of clear yellow—now again a scheme of pure transparency and black.

Gay and pleasing as are the designs turned out the
palette is a small one and invention is circumscribed.
Under such conditions it might well be supposed
that instances of close resemblance between different
species would be numerous, and this in effect is what
we find.

As in Asia with its Euploeines and Danaines,
and in Africa with its Danaines and Acraeines, so
in S. America are the fashions set by two dominant
groups of models. These are the Heliconinae and
the Ithomiinae, both peculiar to this region and both
characterised, like the Old-world Danaids, by slow
flight and great tenacity of life. Both live on poisonous
plants—the Heliconines on Passifloras and the Ithomiines on Solanaceae. In both groups, but more
especially in the Ithomiinae, the species are numerous,
and the number of individuals in a species often
beyond computation. From the point of view of
mimicry these two groups have so much in common
that they may conveniently be considered together.

It was from among the Ithomiines, as already
pointed out, that the models came for the Pierine
mimics of the genus *Dismorphia* upon which Bates
founded the theory of mimicry. Though the Pierine
mimics are the most striking the Heliconines and
Ithomiines are mimicked by members of other groups.
A few Papilios (Pl. X, fig. 8), certain Nymphalines
such as *Protogonius* (Pl. X, fig. 9), *Eresia, Phyciodes*
and *Colaenis* (Pl. XI, fig. 4), together with various
day-flying moths, more particularly of the genera

Castnia and *Pericopis*, are among the well-known mimics of this group of models. The models themselves are very variable in appearance. In one locality the predominant pattern is black with a warm red-brown diagonal bar occupying rather more than a third of the fore wing (Pl. XV, fig. 5), in another it consists of parallel bands of black and fulvous brown with clear yellow patches at the tips of the fore wings (cf. Pl. X, fig. 7), while in yet another locality it is different again. Different localities often have their own peculiar pattern and this affects the various mimics as well as the Ithomiine and Heliconine models.

These groups of different species, some belonging to palatable and some to unpalatable groups, all exhibiting a close resemblance in colour and pattern, are far more strikingly developed in S. America than in either Asia or Africa, and it is not uncommon for eight or ten species to enter into such an association. A group of this sort which possesses unusual interest is the so-called "Transparency Group" from certain parts of the Amazon region. It was originally described by Bates with seven species belonging to six different genera. To-day it is said that no less than 28 species of this peculiar facies are known, though some are excessively rare. The majority are Itho-miines, but two species of the Danaine genus *Ituna,* the Pierine *Dismorphia orise* (Pl. XII, fig. 2), the Swallow-tail *Papilio hahneli,* and several species of diurnal moths belonging to different families (cf. Pl. XII, fig. 4) also enter into the combination.

In connection with it there is a feature of peculiar interest in that the transparent effect is not always produced in the same way. In the Ithomiines such as *Thyridia,* where there are normally two kinds of scales, the wider ones for the most part lose their pigment, become much reduced in size and take on the shape of a stumpy V (Pl. XIV, fig. 3). Also they stand out for the most part more or less at right angles to the wing[1], and the neck by which they are joined to the wing membrane is very short. The longer and narrow form of scales also tend to lose their pigment and become reduced to fine hairs. In *Dismorphia* the scales, which are of one sort, are also reduced in size though apparently not in number. Like the wider scales of the *Thyridia* they tend sometimes to project at right angles to the wing membrane, though not to the same extent as in the Ithomiine : possibly because the neck of the scale is not so short. As in *Thyridia* these reduced scales lose their pigment except in the transition region round the borders of the transparent patches. In *Ituna* there is a difference. The scales are not reduced to the same extent in point of size. Their necks are longer as in normal scales and they lie flat on the wing membrane. The majority of the scales, as in the preceding cases, lose their pigment, but mixed up with them is a certain proportion, about one-quarter,

---

[1] These descriptions are taken from preserved specimens which I owe for the most part to the kindness of Dr Jordan. I have not had an opportunity of examining fresh ones.

in which the pigment is retained. In *Castnia* and in *Anthomysa* the scales on the transparent parts which are without pigment are also somewhat reduced in size, being stumpier than the normal ones. At the same time they tend to stand out at right angles to the wing membrane[1]. The neck here again is shorter in the transparent than in the pigmented scales. A good deal of stress has been laid upon this case by some supporters of the theory of mimicry, since it is supposed to shew that a similar effect can be brought about in a variety of ways; consequently the existence of this assembly of similar transparent forms belonging to various families cannot be put down as due to the effect of similar conditions, but must be regarded as having arisen in each instance in a different manner through the independent action of natural selection[2]. It is doubtful, however, whether such a conclusion necessarily follows from the facts. In all of the cases the process would appear to be similar: loss of pigment, reduction in the size of the scales, and eventually a tendency for the scales to stand at right angles to the wing—this last part of the process apparently depending upon the reduction of the neck of the scale. It has been said that greater transparency is brought about by the scales standing out at right angles in this way, but as the scales them-

---

[1] This is more marked in *Castnia* than in *Anthomysa*. It appears to be a peculiarity of many members of the genus *Castnia* that the scales do not lie so tight as generally in moths. Owing to this, some of the large whole-coloured species have a somewhat fluffy look.

[2] Cf. Poulton, *Essays on Evolution*, 1908, pp. 264-6.

selves are already transparent there would appear
to be no reason why this should be so. Of course
the process has not proceeded in all of the forms
to the same extent. There is least change in *Ituna*
where the scales are not much reduced in size and
where a fair proportion are still pigmented. There
is probably most in an Ithomiine such as *Thyridia*,
where the scales are not only small and entirely without
pigment, but also are for the most part neckless so
that they stand out at right angles to the wing. Having
regard to the fact that several widely separate genera
with different types of scaling formed the starting
points, the final results do not seem to preclude the
supposition that the transparency has arisen through
a similar process in all of them.

It is somewhat remarkable that no Satyrine exhibits
mimicry in S. America, in spite of the fact that transparency of the wings, as in so many of the butterflies
of this region, is quite common in the group. On
the other hand the relatively large number of more
or less mimetic Pierines is a striking feature of S.
America. For the most part they belong to the
genera *Dismorphia* and *Perrhybris*, and resemble the
yellow, black, and brown Heliconines and Ithomiines,
though some of the former genus are mimics of the
small transparent Ithomiines. Some of the species
of *Pereute* with their dark ground colour and the
bright red bar across the fore wing (Pl. XI, fig. 6)
resemble *Heliconius melpomene*, as also does *Papilio
euterpinus*. But some of the most interesting Pierine

mimics are several forms belonging to the genus *Archonias* (Pl. XI, fig. 10) which exhibit the simple and striking arrangement of black, red and white so characteristic of the Swallow-tail Poison-eaters of S. America. They form one of the rare instances of a Pharmacophagus Papilio being mimicked by a butterfly which does not belong to the Swallow-tail group.

As everywhere in the tropics the Papilios of S. America supply a goodly proportion of the mimicry cases. A few, such as *P. zagreus* (Pl. X, fig. 8), enter into the black-brown and yellow Ithomiine-Heliconine combination ; *P. euterpinus* resembles *Heliconius melpomene* (Pl. XI, fig. 5) ; *P. pausanias* is like *Heliconius sulphurea* (Pl. XI, figs. 1 and 2). But this practically exhausts the list of Papilios which mimic Heliconines and Ithomiines. The great majority of mimicking Swallow-tails in S. America find their models among the Poison-eaters of their own family, offering in this respect a contrast to those of Asia where the majority of models are among the Danaines and Euploeines, and of Africa where they are exclusively Acraeines or Danaines.

The Poison-eaters of S. America fall into two well-marked groups which we may call the red-spotted and the dark green groups respectively. The red spotted group form a remarkably compact and uniform assemblage. The general ground colour is a deep black-brown (Pl. XI, figs. 8 and 9), the hind wings are almost invariably marked with red near the centre or towards the outer margin, and the fore wing may or may not bear a patch which is generally whitish in the female, though often of a brilliant blue or green in the male. This simple colour scheme with variations runs throughout about three-quarters (some 40 species) of the Poison-eaters. The same general colour scheme is also found in about two dozen species of the unprotected Swallow-tails. As the total number of the unprotected species is placed by Seitz at less than 100 this means that fully one-quarter of them fall into the general colour scheme adopted by the majority of the Poison-eaters. In many cases the resemblance between mimic and model is so close as to have deceived the most expert entomologists before the structural differences between the groups had been appreciated (cf. Appendix II). The matter is further complicated by the fact that polymorphism is not uncommon, especially among the females of the mimetic forms. *Papilio lysithous* for instance has no less than six distinct forms of female, which differ chiefly in the extent and arrangement of the white markings on the wings, one form lacking them entirely. Several of these forms may occur together in a given locality, and may resemble as many distinct species of Poison-eaters. Thus the three forms *lysithous*, with white on both wings, *rurik*, with white on the fore wing only, and *pomponius* without any white, all fly together in Rio Grande do Sul and respectively mimic the three distinct Pharmacophagus species *nephalion*, *chamissonia*, and *perrhebus* (Pl. XIII). It is worthy of note that mimics are provided by both unprotected groups of Swallow-tails in S. America, whereas in Asia the Cosmodesmus division never provides mimics for Pharmacophagus models (cf. Appendix II).

In the second and smaller group of the Pharmacophagus Swallow-tails the general colour scheme is a more or less dark metallic blue-green with a tendency towards the obliteration of light markings. Some idea of their appearance may be got from the figure of the Central and N. American *P. philenor* on Pl. XVI, fig. 1. Though one or two unprotected Papilios in S. America fall more or less into this colour scheme, the group, from the point of view of mimicry, is not nearly so important as the red-spotted one.

Nevertheless the blue-green Pharmacophagus group as represented by *P. philenor* is supposed to play a considerable part in mimicry in N. America. *P. philenor* is found throughout the greater part of the Eastern United States, straggling up as far as the Canadian border. On the west it is also found reaching up to North California. Over considerable parts of its range are three other Swallow-tails, belonging to the unprotected Papilios, which are regarded by Professor Poulton and others as mimics of *philenor*[1]. One of these, *P. troilus*, is dark brown with a dusting of blue scales over the hind wing (Pl. XVI, fig. 2). The sexes here are more or less alike. *Troilus* stretches up into North-west Canada some way beyond the limits reached by its model. *P. glaucus* is a black and yellow Swallow-tail with two forms of female.

---

[1] Cf. Poulton, *Darwin and the 'Origin,'* 1909, pp. 177–186.

One of these resembles the male while the other is darker and is said to mimic *philenor*. It is known as the *turnus* form and is found more commonly in the southern part of the range of the species, *i.e.* in the country where *philenor* is more plentiful. The third species, *P. asterius*, has a more southerly distribution. Its female is darker and nearer to *philenor* than the male. It must, however, be admitted that none of the three species bears a very close resemblance to *philenor*. It is suggested that this is because *P. philenor* is a tropical form which has only recently invaded N. America. The crossing of *philenor* has, as it were, induced the three mimicking *Papilios* to turn dark, but the model has not been long enough in contact with them for the likeness to become a close one. The explanation, however, hardly accounts for the fact that the best mimic of the three, *P. troilus*, in which both sexes are dark, is found far north of *philenor*. Either the dark colour was established without the influence of the Pharmacophagus model, or else the species rapidly extended its range northwards after having been modified under the influence of *philenor* in the south. But in that case the critic may ask why it does not revert to the original pattern now that it has got beyond the model's sphere of influence. On the whole it seems at present quite doubtful whether any relation of a mimetic nature exists between *P. philenor* and these three species of *Papilio*.

P. philenor is also regarded as serving as a model

for two Nymphaline butterflies in the United States. One of these is the large Fritillary *Argynnis diana* of which the dark female has a markedly blue tint on the upper surface (Pl. XVI, fig. 3). The other is a *Limenitis*[1] related to our own White Admiral. This form, *L. astyanax* (Pl. XVI, fig. 5), is a dark form with a bluish iridescence on the upper surface. It is found, like *P. philenor*, over the greater part of the Eastern States, while to the north, near the Canadian boundary, its place is taken by *L. arthemis* with prominent white bar across both wings (Pl. XVI, fig. 4). There is reason for believing that where the two overlap there is occasional inbreeding, and that the hybrid is the form known as *proserpina*, resembling *astyanax* more than *arthemis*. It must be admitted that in general appearance *L. astyanax* and *Argynnis diana* are more like *Papilio troilus* than *P. philenor*. In explanation it has been suggested that all the mimics are on the way to resembling *P. philenor*, and consequently we should expect them at certain stages to shew more resemblance to one another than to the form they have all as it were set out to mimic. On this view they will all arrive at a close resemblance to *philenor* in time. Another explanation is that favoured by Professor Poulton on which it is assumed that we are here dealing with a case of Müllerian Mimicry, all of the species in question being distasteful with the exception perhaps of *A. diana*. Thus *troilus* and *astyanax* though distasteful are less so than

[1] The N. American members of this genus are often referred to as *Basilarchia*.

philenor. Hence it is of advantage to them to have even a chance of being mistaken for the more obnoxious *philenor*, and so the one has come from the black and yellow Swallow-tail pattern and the other from the white-banded *arthemis* form to what they are, *i.e.* more alike to one another than to *philenor*. They now form a Müllerian combination for mutual protection along with the dark females of *glaucus* and *asterius*. But they are themselves still moderately distasteful so that it is to the advantage of the female of *Argynnis diana* to mimic them. Whether they are all on the way to resembling *philenor* more closely, or whether they have sufficiently vindicated their inedible properties and are now stationary, it is for the future to reveal to posterity. Lastly we have the view that these different species have attained their present coloration entirely independently of one another, and that we are not here concerned with mimicry at all. Since the sole evidence available at present is that based on general appearance and geographical distribution, the view taken of this case must rest largely upon personal inclination.

Though the cases just quoted are only very problematically mimetic, N. America has yet several examples of resemblance between distantly related forms as close as any that occur in the tropics. In this region are found two species of the genus *Danais*— *D. archippus* occurring all over the United States and reaching up northwards into Canada, *D. berenice* found in the South-eastern States, *e.g.* in Florida, where it is said to be more abundant than *archippus*.

D. archippus (Pl. XVI, fig. 8) is very similar to the oriental *D. plexippus* (Pl. IV, fig. 2), from which perhaps its most notable difference lies in the extent and arrangement of the white spots near the tip of the fore wing. *D. berenice* is not unlike *archippus* in its general colour scheme but is smaller and darker (Pl. XVI, fig. 9).

We have already had occasion to mention the common Nymphaline, *Limenitis arthemis* (Pl. XVI, fig. 4) which is found in Canada and the Northeastern States. Widely spread over N. America is a close ally of this species, *L. archippus*, which, though so similar in structure and habits, is very different in external appearance. As appears from Pl. XVI, fig. 6, *L. archippus* is remarkably like the Danaid which bears the same specific name. In the Southern States *L. archippus* is replaced by a form slightly different in details of pattern and distinctly darker, *L. floridensis* (=*eros*) (Pl. XVI, fig. 7). In Florida occurs also the darker N. American Danaid, *D. berenice*, to which the colour of *L. floridensis* approximates more than to *D. archippus*, and it is of interest that although the last named is also found in this locality it is said to be much less abundant than *D. berenice*. Nevertheless it appears to be true that the range of *L. floridensis* is much more extensive than that of its model; in other words, that there are considerable regions where *L. floridensis* and *D. archippus* coexist, and from which *L. archippus* and *D. berenice* are wanting.

CHAPTER V

SOME CRITICISMS

THE facts related in the last two chapters are sufficient to make it clear that these remarkable resemblances between species belonging as a rule to widely different groups constitute a real phenomenon, and as such demand an explanation. One explanation, that in terms of the theory of mimicry, has already been outlined, and we may now turn to consider it in more detail. Some years ago Wallace[1], combating the suggestion that these instances of resemblance might be mere coincidences, laid down five conditions which he stated were applicable to all such cases, and rendered utterly inadequate any explanation other than in terms of natural selection. These five conditions are of historical interest and may also serve as a peg for sundry criticisms in connection with the mimicry theory. They are as follows :

(1) That the imitative species occur in the same area and occupy the very same station as the imitated.

(2) That the imitators are always the more defenceless.

[1] *Darwinism,* 1890 (1st Edition 1889), p. 264.

(3) That the imitators are always less numerous in individuals.

(4) That the imitators differ from the bulk of their allies.

(5) That the imitation, however minute, is *external* and *visible* only, never extending to internal characters or to such as do not affect the external appearance.

In offering certain criticisms of the mimicry explanation it will be convenient to do so in connection with these five conditions which Wallace regarded as constant for all cases of mimetic resemblance.

(1) *That the imitative species occur in the same area and occupy the very same station as the imitated.*

This on the whole is generally true. It is well shewn in some of the most striking cases such as those of the Old-World Papilios that mimic Danaines, or of the Dismorphias and their Ithomiine models. In many of these cases the range of neither model nor mimic is a very wide one, yet the mimic is found strictly inside the area inhabited by the model. *Papilio agestor*, for instance, is only found where *Caduga tytia* occurs, nor is *P. mendax* known outside the area frequented by *Euploea rhadamanthus*. Even more striking in this respect are some of the Ithomiine-Dismorphia resemblances in the New World. The Ithomiine models are as a rule very local though very abundant. Two hundred miles away the predominant Ithomiine often bears quite a distinct pattern, and when this is the case the mimicking *Dismorphia* is generally changed in the same sense.

4—2

But though mimic and model may be found together in the same locality, they do not always occupy the same station in the sense that they fly together. According to Seitz[1] the Dismorphias themselves do not fly with the Ithomiines which they mimic. The occurrence of butterflies is largely conditioned by the occurrence of the plants on which the larva feeds, and this is especially true of the female, which, as has already been noticed, is more commonly mimetic than the male. The female of *Papilio polytes*, for instance, is found flying where are to be found the wild citronaceous plants on which its larva feeds. On the other hand, its so-called models, *Papilio hector* and *P. aristolochiae*, are generally in the proximity of the Aristolochias on which their larvae feed. The two plants are not always found together, so that one frequently comes across areas where *P. polytes* is very abundant while the models are scarce or absent.

Though in the great majority of cases the imitator and the imitated occur in the same locality, this is not always so. The female of the Fritillary *Argynnis hyperbius* (Pl. IV, fig. 3), for instance, is exceedingly difficult to distinguish from *Danais plexippus* when flying, although when at rest the difference between the two is sufficiently obvious. Both insects are plentiful in Ceylon but inhabit different stations. The Danaid is a low-country insect, while the Fritillary is not found until several thousand feet up. The two species affect entirely different stations and hardly

[1] *Macrolepidoptera of the World. Fauna Americana*, p. 98.

come into contact with each other. Where one is plentiful the other is not found. It has been suggested that migratory birds may have come into play in such cases. The bird learns in the low country that *D. plexippus* is unpleasant, and when it pays a visit to the hills it takes this experience with it and avoids those females of the Fritillary which recall the unpleasant Danaine.

Migratory birds have also been appealed to in another case where the resembling species are even further removed from one another than in the last case. *Hypolimnas misippus* is common and widely spread over Africa and Indo-Malaya, and the male (Pl. IV, fig. 8) bears a simple and conspicuous pattern—a large white spot bordered with purple on each of the very dark fore and hind wings. The same pattern occurs in the males of two other Nymphalines allied to *H. misippus*, viz. *Athyma punctata* and *Limenitis albomaculata*. The two species, however, have a distribution quite distinct from that of *H. misippus*, being found in China. It has nevertheless been suggested by Professor Poulton[1] that the case may yet be one of mimicry. According to his explanation, *H. misippus* is unpalatable, the well-known association of its female with *Danais chrysippus* being an instance of Müllerian mimicry. Migratory birds did the rest. Having had experience of *H. misippus* in the south, on their arrival in China they spared such specimens of *Athyma punctata* and *Limenitis*

[1] *Essays on Evolution*, 1908, p. 381.

albomaculata as approached most nearly to *H. misippus*
in pattern, and so brought about the resemblance.
The explanation is ingenious, but a simpler view will
probably commend itself to most.   Other cases are
known in which two butterflies bear a close resemblance
in pattern and yet are widely separated geographically.
Several species of the S. American Vanessid genus
Adelpha are in colour scheme like the African *Planema
poggei* which serves as a model for more than one
species.   The little S. American *Phyciodes leucodesma*
would almost certainly be regarded either as a model
for or a mimic of the African *Neptis nemetes*, did the
two occur together.   Nevertheless examples of close
resemblance between butterflies which live in different
parts of the world are relatively rare and serve to
emphasise the fact that the great bulk of these
resemblance cases are found associated in pairs or in
little groups.

(2)   *That the imitators are always the more defenceless.*

In the case of butterflies " defence " as a rule denotes
a disagreeable flavour rendering its possessor distasteful
to birds and perhaps to other would-be devourers.
Feeding experiments with birds (cf. Chapter IX)
suggest that certain groups of butterflies, notably
the Danaines, Acraeines, Heliconines, Ithomiines and
Pharmacophagus Papilios—groups from which models
are generally drawn—are characterised by a disagreeable
taste, while as a rule this is not true for the mimics.
This distasteful quality is frequently accompanied
by a more or less conspicuous type of coloration,

though this is by no means always so. Many Euploeas are sombre inconspicuous forms, and it is only some of the Ithomiines that sport the gay colours with which that group is generally associated. The members of the distasteful groups usually present certain other peculiarities. Their flight is slower, they are less wary, their bodies are far tougher, and they are more tenacious of life. The slow flight is regarded as an adaptation for exhibiting the warning coloration to the best advantage, but from the point of view of utility it is plausible to suggest that the insect would be better off if in addition to its warning coloration it possessed also the power of swift flight[1]. It is possible that the peculiar slowness of flight of these unpalatable groups is necessitated by the peculiar tough but elastic integument which may present an insufficiently firm and resistant skeletal basis for sharp powerful muscular contraction, and so render swift flight impossible. It is stated that the flight of the mimics is like that of the model, and in some cases this is undoubtedly true. But in a great many cases it certainly does not hold good. *Papilio clytia* (Pl. I, figs. 7 and 8) is a strong swift flyer very unlike the Danaine and Euploeine which it is supposed to mimic. The flight of the female of *Hypolimnas misippus* (Pl. IV, fig. 7) is quite distinct from that of *Danais chrysippus*, while the mimetic

---

[1] These "unpalatable" butterflies are sometimes extensively preyed upon by insectivorous birds, when they fall an easier prey owing to their slowness (cf. p. 112).

forms of *P. polytes* fly like the non-mimetic one, a mode of flight so different from that of the two models that there is no difficulty in distinguishing them many yards away. Swift flight must be reckoned as one of the chief modes of defence in a butterfly, and on this score the mimic is often better off than the model. And of course it must not be forgotten that where the mode of flight is distinct the protective value of the resemblance must be very much discounted.

(3) *That the imitators are always less numerous in individuals.*

In the majority of cases this is certainly true. Probably all the Old-World Papilios that mimic Danaines are scarcer, and frequently very much scarcer, than their models. This is very evident from a study of the more comprehensive priced catalogues of Lepidoptera. The mimic is generally a more expensive insect than the model, and not infrequently it costs as many pounds as the model does shillings. But the rule is not universal. *Papilio polytes* is often much more common than either of its models. The remarkable Pierines, *Archonias tereas* and *A. critias* (Pl. XI, fig. 10) as a rule far outnumber the Pharmacophagus Swallow-tail which they mimic. Or again the Chalcosid moth *Callamesia pieridoides*[1] is a more abundant insect than the Bornean Pierine *Delias cathara* which it closely resembles.

It has sometimes been suggested in explanation

---

[1] See Shelford, *Proc. Zool. Soc.* 1902, p. 260. A coloured figure of both species is given in the paper.

of the greater abundance of the mimic that in such cases we are concerned with Müllerian mimicry, that since both of the species concerned are distasteful there is not, strictly speaking, either a mimic or a model, and consequently the relative proportions have not the significance that they possess where the mimicry is of the simple Batesian type. It is, however, very doubtful whether such an explanation is of any value, for, as will appear later, there are grave objections to accepting the current theory as to the way in which a resemblance is established on Müllerian lines (cf. pp. 72–74).

(4) *That the imitators differ from the bulk of their allies.*

What importance we attach to this condition must depend upon our interpretation of the word "allies"—whether, for example, we use it for a small group of closely connected species, for a genus, for a group of genera, or in an even wider sense. Perhaps an example will serve to make the difficulty more clear. As already noticed, the S. American genus *Dismorphia* belongs to the family of Pieridae or "whites." Also certain species of *Dismorphia* bear a close resemblance to certain species of Ithomiines, a noteworthy example being *D. praxinoe* and *Mechanitis saturata* (Pl. X, figs. 3 and 7), in which the pattern, colour, and shape of the two species are all far removed from what is usually understood by a "white." It must not be forgotten, however, that these matters are generally discussed by European

naturalists who have grown up in a region where
the majority of the "whites" are more or less white.
For this reason the statement that *D. praxinoe* differs
from the bulk of its allies is likely to meet with
general acceptance, especially as some of the species
of the genus itself (e.g., *D. cretacea*, Pl. X, fig. 1)
are regular whites in appearance. But when we
come to look at the genus *Dismorphia* as a whole the
matter assumes another complexion. Seitz [1] recognises
75 species of which about a dozen are predominantly
white. The rest present a wonderful diversity of
colour and pattern. Black predominates on the fore
wings, and the insect is frequently marked with gay
patches of yellow, bright brown, scarlet, or blue.
Forms which from their colour are clearly not mimics
present nevertheless the general pattern and shape
of other forms which bear a strong resemblance to
some Ithomiine. Sometimes a change of colour in
certain patches from blue or yellow to bright brown
would make all the difference between a non-imitative
and an imitative species. Moreover, the non-imitative
forms frequently have the peculiar narrow wing, so
unusual in a Pierine, which enhances the resemblance
of the mimicking species to the Ithomiine model,
and which to some extent occurs even in *D. cretacea*.
Clearly we are not justified in saying that *D. praxinoe*
differs from the bulk of its allies, for inside the genus
there are many non-imitative species which differ

[1] *Macrolepidoptera of the World. Fauna Americana*, pp. 98–104,
Plates 28–30.

from it in some particulars and are alike it in others. There is a distinct family resemblance among the bulk of the Dismorphias, including practically all the mimetic forms, and on the whole the resemblances between the imitative and the non-imitative forms are as noteworthy as the differences. Though not exhibited in so striking a fashion, the same is to a large extent true of a large proportion of the cases of mimicry. It is on the whole unusual to find cases where a single species departs widely from the pattern scheme of the other members of the genus and at the same time resembles an unrelated species. Two of the best instances are perhaps those of *Limenitis archippus* (p. 49) and of the Pierid *Pareronia* (p. 23). Of the total number of mimicry instances a high proportion is supplied by relatively few groups. In each region several main series of models and mimics run as it were parallel to one another. In Asia, for example, we have the Papilio-Danaine series where the colour-patterns of a series of Danaines, all nearly related, are closely paralleled by those of a section of the genus *Papilio*, and by those of the Satyrid genus *Elymnias*. In Africa there is a similar Papilio-Danaine series though of less extent. Africa has a group of models not found in Asia, and the Papilio-Danaine series is as it were curtailed by the Papilio-Planema series with which to some extent runs parallel the genus *Pseudacraea*. These phenomena of parallel series have been mentioned here as shewing that mimicry tends to run in certain groups and that in many cases at any rate little meaning can be attached to the statement that the imitators differ from the bulk of their allies.

The fifth of Wallace's conditions is clear and needs no discussion.

It is evident that at any rate a large proportion of the instances of close resemblance do not fulfil all of the conditions laid down by Wallace. Nevertheless we should expect them to do so if the resemblance has been brought about by the cumulative effect of natural selection on small favourable variations. Clearly there is a *prima facie* case for doubting whether we must of necessity ascribe all resemblance of the kind to natural selection, and in the next few chapters we shall discuss it in more detail from several points of view.

CHAPTER VI

"MIMICRY RINGS"

HAVING reviewed briefly some of the most striking phenomena of what has been termed mimicry, we may now inquire whether there are good grounds for supposing that these resemblances have been brought about through the operation of natural selection or whether they are due to some other cause. If we propose to offer an explanation in terms of natural selection we are thereby committed to the view that these resemblances are of the nature of adaptation. For unless we grant this we cannot suppose that natural selection has had anything to do either with their origin or with their survival. Granting then for the present the adaptational nature of these mimetic resemblances, we may attempt to deduce from them what we can as to the mode of operation of natural selection. In doing so we shall bear in mind what may be called the two extreme views : viz. (a) that the resemblance has been brought about through the gradual accumulation of very numerous small variations in the right direction through the operation of natural selection, and (b) that the mimetic form came into being as a sudden sport or mutation, and that natural selection is responsible merely for its survival and the elimination of the less favoured form from which it sprang.

There is a serious difficulty in the way of accepting the former of these two views. If our two species, model and would-be mimic are, to begin with, markedly different in pattern, how can we suppose that a slight variation in the direction of the model on the part of the latter would be of any value to it? Take for example a well-known South American case—the resemblance between the yellow, black, and brown Ithomiine, *Mechanitis saturata* (Pl. X, fig. 7) and the Pierine, *Dismorphia praxinoe* (Pl. X, fig. 3). The latter belongs to the family of the "whites," and entomologists consider that in all probability its ancestral garb was white with a little black like the closely allied species *D. cretacea* (Pl. X, fig. 1). Can we suppose that in such a case a small development of brown and black on the wings would be sufficient to recall the Ithomiine and so be of service to the *Dismorphia* which possessed it? Such a relatively slight approach to the Ithomiine colouring is shewn by the males of certain South American "whites" belonging to the genus *Perrhybris* (Pl. X, figs. 4 and 5). But the colour is confined to the under-surface and the butterflies possessing it could hardly be confused with a *Mechanitis* more than their white relations which entirely lack such a patch of colour. If birds regarded white butterflies as edible it is difficult to suppose that they would be checked in their attacks by a trifling patch of colour while the main ground of the insect was still white. But unless they avoided those with the small colour patch there would be an end of natural selection in so far as the patch was concerned, and it would have no opportunity of developing further through the operation of that factor. This is the difficulty of the initial variation which has been clearly recognised by most of the best known supporters of the theory of mimicry. Bates himself offered no suggestion as to the way in which such a form as a Pierid could be conceived of as beginning to resemble an Ithomiine[1]. Wallace supposed that the Ithomiines were to start with not so distinct from many of the edible forms as they are to-day, and that some of the Pierines inhabiting the same district happened to be sufficiently like some of the unpalatable forms to be mistaken for them occasionally[2].

The difficulty of the initial variation had also occurred to Darwin, and he discusses it in an interesting passage which is so important that we may quote it here in full :

It should be observed that the process of imitation probably never commenced between forms widely dissimilar in colour. But starting with species already somewhat like each other, the closest resemblance, if beneficial, could readily be gained by the above means ; and if the imitated form was subsequently and gradually

[1] "In what way our *Leptalis* ($=Dismorphia$) originally acquired the general form and colour of Ithomiae I must leave undiscovered." *Trans. Linn. Soc.* vol. 23, 1862, p. 513.

[2] *Darwinism*, 1890, pp. 242–244.

modified through any agency, the imitating form would be led
along the same track, and thus be altered to almost any extent, so
that it might ultimately assume an appearance or colouring wholly
unlike that of the other members of the family to which it belonged.
There is, however, some difficulty on this head, for it is necessary
to suppose in some cases that ancient members belonging to several
distinct groups, before they had diverged to their present extent,
accidentally resembled a member of another and protected group
in a sufficient degree to afford some slight protection; this having
given the basis for the subsequent acquisition of the most perfect
resemblance[1].

Both Darwin and Wallace recognised clearly this
difficulty of the initial variations, and both suggested
a means of getting over it on similar lines. Both
supposed that in general colour and pattern the groups
to which model and mimic belonged were far more
alike originally than they are to-day. They were
in fact so much alike that comparatively small variations in a favourable direction on the part of the mimic
would lead to its being confused with the unpalatable
model. Then as the model became more and more
conspicuously coloured, as it developed a more and
more striking pattern warning would-be enemies of
its unpleasant taste, the mimic gradually kept pace
with it through the operation of natural selection,
in the shape of the discriminating enemy, eliminating
those most unlike the model. The mimic travelled closely
in the wake of the model, coaxed as it were by natural
selection, till at last it was far removed in general
appearance from the great majority of its near relations.

---

[1] *Origin of Species*, 6th Edition, 1891, p. 354.

In this way was offered a comparatively simple method of getting over the difficulty of applying the principle of natural selection to the initial variations in a mimetic approach on the part of one species to another. But it did not escape Darwin's penetration that such an argument would not always be easy of application—that there might be cases where a given model was mimicked by members of several groups of widely differing ancestral pattern, and that in these cases it would be difficult to conceive of members of each of the several groups shewing simultaneous variations which would render them liable to be mistaken for the protected model. The difficulty may perhaps be best illustrated if we consider a definite case.

It is a feature of mimetic resemblances among butterflies that a given species in a given locality may serve as a model for several other species belonging to unrelated groups. Generally such mimics belong to presumably palatable species, but other presumably unpalatable species may also exhibit a similar coloration and pattern. In this way is formed a combine to which the term "mimicry ring" has sometimes been applied. An excellent example of such a mimicry ring is afforded by certain species of butterflies in Ceylon, and is illustrated on Plate IV. It is made up in the first place of two species belonging to the presumably distasteful Danaine group, viz. *Danais chrysippus* and *D. plexippus*. The latter is a rather darker insect but presents an unmistakable general likeness to *D. chrysippus*. Those who believe in

Müllerian mimicry would regard it as an excellent example of that phenomenon. For those who believe only in Batesian mimicry *D. plexippus*, being the scarcer insect, must be regarded as the mimic and *D. chrysippus* as the model. In both of these species the sexes are similar, whereas in the other three members of the "ring" the female alone exhibits the resemblance. One of these three species is the common Nymphaline, *Hypolimnas misippus*, of which the female bears an extraordinary likeness to *D. chrysippus* when set and pinned out on cork in the ordinary way. The male, however (Pl. IV, fig. 8), is an insect of totally different appearance. The upper surfaces of the wings are velvety black with a large white patch bordered with purple in the middle of each[1]. The "ring" is completed by the females of *Elymnias undularis* and *Argynnis hyperbius*. The former of these belongs to the group of Satyrine butterflies and the female is usually regarded as a mimic of *D. plexippus*, which it is not unlike in so far as the upper surface of the wings is concerned. Here again the male is an insect of totally dissimilar appearance. Except for a border of lighter brown along the outer edges of the hind wings the upper surface is of a uniform deep purple-brown

---

[1] *H. misippus* was at one time regarded as a clear case of Batesian mimicry. But in view of its plentifulness, of the fact that it may be abundant outside the area inhabited by its model, and of the ease with which it can establish itself in parts remote from its original habitat, *e.g.* S. America, it has come to be regarded by certain supporters of the mimicry theory as a Müllerian mimic. Cf. Poulton, *Essays on Evolution*, 1908, pp. 215–217.

all over (Pl. IV, fig. 6). In *Argynnis hyperbius*
the appearance is in general that of the Fritillary
group to which it belongs. But in the female the
outer portion of the fore wings exhibits much black
pigment and is crossed by a broad white band similar
to that found in the same position on the wing of
D. plexippus (Pl. IV, fig. 2).

Of the five species constituting this little "mimicry
ring" in Ceylon two, on the current theory of mimicry,
are to be regarded as definitely unpalatable, one
(*H. misippus*) as doubtfully so, while the Satyrine
and the Fritillary are evidently examples of simple
or Batesian mimicry.

Now such examples as this of simultaneous mimicry
in several species are of peculiar interest for us when
we come to inquire more closely into the process
by which the resemblances can be supposed to have
been brought about. Take for example the case of
E. undularis. The male is evidently an unprotected
insect in so far as mimicry is concerned, while the
female exhibits the general pattern and coloration
characteristic of the warningly coloured and presumably distasteful species *D. plexippus* or *D. chrysippus*. If we are to suppose this to have been brought
about by the operation of natural selection it is clear
that we must regard the colour and pattern of the
male as the original colour and pattern of both sexes.
For natural selection cannot be supposed to have
operated in causing the male to pass from a protected
to an unprotected condition, or even in causing him

to change one unprotected condition for another. Probably all adherents of the mimicry theory would be agreed in regarding the male of *Elymnias undularis* as shewing the ancestral coloration of the species, and in looking upon the female as having been modified to her own advantage in the direction of *D. plexippus*. The question that we have to try to decide is how this has come about—whether by the accumulation of slight variations, or whether by a sudden change or mutation in the pattern and colour of the female by which she came to resemble closely the Danaine. It is clear that if *D. plexippus* were what it is to-day before the mimetic approach on the part of *E. undularis* began, small variations in the latter would have been of no service to it. The difference between the two species would have been far too great for individuals exhibiting slight variation in the direction of *D. plexippus* to stand any chance of being confused with this species. And unless such confusion were possible natural selection could not work. There is, however, an immediate way out of the difficulty. We may suppose that the coloration of the male of the mimic, *E. undularis,* is not only the ancestral colour of its own species but also of the model. *D. plexippus* on this supposition was very like *E. undularis,* of which both sexes were then similar to what the male is to-day. The pattern is, however, an inconspicuous one, and it can be imagined that it might be to the advantage of *D. plexippus* to don a brighter garb for the advertisement of its unpleasant qualities.

Variations in the direction of a more conspicuous pattern would for that reason tend to be preserved by natural selection, until eventually was evolved through its means the well-marked pattern so characteristic of the model to-day. If in the meantime variations in the same direction occurred among the females of *E. undularis* these would tend to be preserved through their resemblance to the developing warning pattern of the distasteful Danaine model. The development of model and mimic would proceed *pari passu,* but if the sexes of the mimic differ, as in this case, we must suppose the starting-point to have been the condition exhibited by the male of the mimicking species.

But *Argynnis hyperbius* is also a species in which the female mimics *D. plexippus*; and by using the same argument as that just detailed for *Elymnias undularis* we can shew that the Danaine model, *D. plexippus,* must also have been like the male of *Argynnis hyperbius.* And if the resemblance of *A. hyperbius* was developed subsequently to that of *E. undularis,* then both *D. plexippus* and *E. undularis* must at one time have been like the male of *A. hyperbius,* a proposition to which few entomologists are likely to assent. Further, since the female of *H. misippus* also comes into the *plexippus-chrysippus* combine we must suppose that these species must at some time or another have passed through a pattern stage like that of the *misippus* male.

It is scarcely necessary to pursue this argument

further, for even the most devoted adherents of the
theory of mimicry as brought about by the operation
of natural selection on small variations are hardly
likely to subscribe to the phylogenetic consequences
which it must entail in cases where a model is mimicked
by the females of several species whose males are
widely dissimilar in appearance.

Even if we suppose the two Danaines to have
been originally like the male of one of the three mimics,
we must still suppose that the females of the other
two originated as "sports," sufficiently near to Danaines
to be confused with them. But if such sports can
be produced suddenly by some mutational process
not at present understood, why should not these
sports be the females of the three mimicking species
as we see them at present ? Why need we suppose
that there were intermediate stages between the
mimicking female and the original hypothetical female
which was like the male ? If a sport occurred which
was sufficiently similar to an unpalatable species to
be confused with it, it is theoretically demonstrable
that, although relatively scarce to start with, it would
rapidly increase at the expense of the unprotected
male-like female until the latter was eliminated. We
shall, however, return in a later chapter (p. 96) to
the argument by which this view can be supported.

So far we have discussed what we called the two
extreme views as to the way in which a mimetic
resemblance may be supposed to have originated. Of
the two that which assumes the resemblance to have

been brought about by a succession of slight variations must also assume that model and mimic were
closely alike to start with, and this certainly cannot
be true in many cases. On the other hand, there is
so far no reason against the idea of supposing the
resemblance to have originated suddenly except what
to most minds will probably appear its inherent im-
probability.

There are writers on these questions of mimicry
who adopt a view more or less intermediate between
those just discussed. They regard the resemblance as
having arisen in the first place as a sport of some
magnitude on the part of the mimic, rendering it
sufficiently like the model to cause some confusion
between the two. A rough-hewn resemblance is first
brought about by a process of mutation. Natural
selection is in this way given something to work on,
and forthwith proceeds to polish up the resemblance
until it becomes exceedingly close. Natural selection
does not originate the likeness, but, as soon as a rough
one has made its appearance, it comes into operation
and works it up through intermediate stages into the
finished portrait. It still plays some part in the
formation of a mimetic resemblance though its rôle is
now restricted to the putting on of the finishing touches.
Those who take this view hold also that the continued
action of natural selection is necessary in order to keep
the likeness up to the mark. They suppose that if
selection ceases the likeness gradually deteriorates
owing to the coming into operation of a mysterious

process called regression. This idea involves certain conceptions as to the nature of variation which we shall discuss later.

Though it is difficult to regard Batesian mimicry as produced by the accumulation of small variations through natural selection, it is perhaps rather more plausible to suppose that such a process may happen in connection with the numerous instances of Müllerian mimicry. For since the end result is theoretically to the advantage of both species instead of but one, it is possible to argue that the process would be simplified by their meeting one another halfway, as Müller[1] himself originally suggested. Variations on the part of each in the direction of the other would be favourably selected, the mimicry being reciprocal.

Difficulties, however, begin to arise when we inquire into the way in which this unification of pattern may be conceived of as having come about. By no one have these difficulties been more forcibly presented than by Marshall[2] in an able paper published a few years ago, and perhaps the best way of appreciating them is to take a hypothetical case used by him as an illustration.

Let us suppose that in the same area live two equally distasteful species A and B, each with a conspicuous though distinct warning pattern, and each sacrificing 1000 individuals yearly to the education of young

[1] An English translation of Müller's paper is given by Meldola, *Proc. Ent. Soc.*, 1879, p. xx.

[2] *Trans. Ent. Soc. Lond.*, 1908, p. 93.

birds. Further let it be supposed that A is a common species of which there are 100,000 individuals in the given area, while B is much rarer, and is represented by 5000. The toll exacted by young birds falls relatively more lightly upon A than upon B, for A loses only 1 %, whereas B's loss is 20 %. Clearly if some members of B varied so that they could be mistaken for A it would be greatly to their advantage, since they would pass from a population in which the destruction by young birds was 20 % to one in which it would now be rather less than 1 %. Moreover, as the proportion of B resembling A gradually increased owing to this advantage, the losses suffered by those exhibiting the original B pattern would be relatively heavier and heavier until the form was ultimately eliminated. In other words, it is theoretically conceivable that of two distasteful species with different patterns the rarer could be brought to resemble the more abundant.

We may consider now what would happen in the converse case in which the more numerous species exhibited a variation owing to which it was confused with the rarer. Suppose that of the 100,000 individuals of A 10,000 shewed a variation which led to their being mistaken for B, so that there are 90,000 of the A pattern and 15,000 of the B pattern of which 10,000 belong to species A. A will now lose 1000 out of the 90,000 having the A pattern, and $\frac{2}{3} \times 1000$ out of the 10,000 of species A which exhibit the B pattern. The toll of the birds will be $\frac{1}{90}$ of those keeping the original A pattern, and $\frac{2}{30}$ of those of species A which have assumed the B pattern. The mortality among the mimetic members of A is six times as great as among those which retain the type form. It is clear therefore that a variation of A which can be mistaken for B is at a great disadvantage as compared with the type form[1], and consequently it must be supposed that the Müllerian factor, as the destruction due to experimental tasting by young birds is termed, cannot bring about a resemblance on the part of a more numerous to a less numerous species. Further, as Marshall goes on to shew, there can be no approach of one species to the other when the numbers are approximately equal. A condition essential for the establishing of a mimetic resemblance on Müllerian lines, no less than on Batesian, is that the less numerous species should take on the pattern of the more numerous. Consequently the argument brought forward in the earlier part of this chapter against the establishing of such a likeness by a long series of slight variations is equally valid for Müllerian mimicry[2].

[1] Provided of course that the type form remains in the majority. If the variation occurred simultaneously in more than 50 % of A the advantage would naturally be with the variation.

[2] It is possible to imagine an exceptional case though most unlikely that it would occur. Suppose for example that there were a number of distasteful species, say 20, all of different patterns, and suppose that in all of them a particular variation occurred simultaneously; then if the total shewing that variation from among the 20 species were greater than the number of any one of the species, all of the 20 species would come to take on the form of the new variation. In this way it is imaginable that the new pattern would gradually engulf all the old ones.

CHAPTER VII

THE CASE OF *PAPILIO POLYTES*

MANY instances of mimicry are known to-day, but comparatively few of them have been studied in any detail. Yet a single carefully analysed case is worth dozens which are merely superficially recorded. In trying to arrive at some conception of the way in which the resemblance has come about we want to know the nature and extent of the likeness in the living as well as in the dead; the relative abundance of model and mimic; what are likely enemies and whether they could be supposed to select in the way required, whether the model is distasteful to them; whether intermediate forms occur among the mimics; how the various forms behave when bred together, etc., etc. Probably the form that from these many points of view has, up to the present, been studied with most care is that of the Swallow-tail, *Papilio polytes*. It is a common butterfly throughout the greater part of India and Ceylon, and closely allied forms, probably to be reckoned in the same species, reach eastwards through China as far as Hongkong. *P. polytes* is one of those species which exhibit polymorphism in the female sex. Three distinct forms of female are known, of which one is like the male, while the other two are very different. Indeed for many years they were regarded as distinct species, and given definite specific names.   To Wallace belongs the credit of shewing that these three forms of female are all to be regarded as wives of the same type of male[1].   He shewed that there were no males corresponding to two of the females; also that the same one male form was always to be found wherever any of the females occurred.   As the result of breeding experiments in more recent years Wallace's conclusions have been shewn to be perfectly sound.

The male of *polytes* (Pl. V, fig. 1) is a handsome blackish insect with a wing expanse of about four inches.   With the exception of some yellowish-white spots along their outer margin the fore wings are entirely dark.   Similar spots occur along the margin of the hind wing also, while across the middle runs a series of six yellowish-white patches producing the appearance of a broad light band.   The thorax and abdomen are full black, though the black of the head is relieved by a few lighter yellowish scales.   The under surface is much like the upper, the chief difference being a series of small and slightly reddish lunules running outside the light band near the margin of the hind wing (Pl. V, fig. 1 *a*).   In some specimens these markings are almost absent.   One form of female is almost exactly like the male (Pl. V, fig. 2), the one slight difference being that the lunules on the under surface of the hind wing are generally a trifle larger.   For brevity she may be called the *M* form.   The second form of female

[1] *Trans. Linn. Soc.* vol. 24, 1866.

differs in many respects from the male and the M female. Instead of being quite dark, the fore wings are marked by darker ribbed lines on a lighter ground[1] (Pl. V, fig. 3). The hind wings shew several marked differences from those of the male. Of the series of six patches forming the cross band the outermost has nearly disappeared, and the innermost has become smaller and reddish. The middle four, on the other hand, have become deeper, reaching up towards the insertion of the wing, and are pure white. A series of red lunules occurs on the upper surface outside the white band, and the yellowish-white marginal markings tend to become red. These differences are equally well marked on the under surface (Pl. V, fig. 3 a). The colour of the body, however, remains as in the male. From the resemblance shewn by this form to another species of Swallow-tail, *Papilio aristolochiae* (Pl. V, fig. 5), we shall speak of it as the A form.

The third form of female is again very distinct from the other two. The fore wings are dark but are broken by an irregular white band running across the middle (Pl. V, fig. 4), and there is also an irregular white patch nearer the tips of the wing. The hind wings, on the other hand, are characterised by having only red markings. The yellowish-white band of the male is much reduced and is entirely red, while the red lunules are much larger than in the A form. The under surface (Pl. V, fig. 4 a) corresponds closely with the

---

[1] These darker ribs are also present in the male and M female but are obscured owing to the generally deeper colour.

upper. The body remains black as in all the other forms. This type of female bears a resemblance to *Papilio hector* (Pl. V, fig. 6), and for that reason we shall speak of it as the *H* form. It should be added that these three forms of female are quite indistinguishable in the larval and chrysalis stages.

It was Wallace who first offered an explanation of this interesting case in terms of mimicry. According to this interpretation *P. polytes* is a palatable form. The larva, which feeds on citronaceous plants, and the chrysalis are both inconspicuous in their natural surroundings. They may be regarded as protectively coloured, and consequently edible and liable to persecution. The original coloration is that of the male and the *M* female. From this the other two forms of female have diverged in the direction of greater instead of less conspicuousness, although the presumed edibility of the insect might have led us to think that a less conspicuous coloration would have been more to its advantage. But these two females resemble the two species *Papilio aristolochiae* and *Papilio hector*, which, though placed in the same genus as *P. polytes*, belong to a very different section of it[1]. The larvae of these two species are conspicuously coloured black and red with spiny tubercles. They feed upon the poisonous *Aristolochia* plants. For these reasons and also from the fact that the butterflies themselves are both conspicuous and plentiful it is inferred that they are unpalatable. In short, they are the models upon

[1] See Appendix II, p. 158.

which the two *polytes* females that are unlike the male
have been built up by natural selection.

The suggestion of mimicry in this case is supported
by the fact that there is a general correspondence
between the areas of distribution of model and mimic.
P. hector is not found outside India and Ceylon, and
the *H* female of *P. polytes* is also confined to this area.
P. aristolochiae, on the other hand, has a much wider
range, almost as wide indeed as that of *P. polytes*
itself. Generally speaking the *A* female accompanies
P. aristolochiae wherever the latter species is found.
Beyond the range of *P. aristolochiae*, in northern China,
the *M* female alone is said to occur. On the other
hand, as the matter comes to be more closely studied
exceptions are beginning to turn up. The *H* female,
for instance, is found on the lower slopes of the Hima-
layas, far north of the range of *P. hector*, and there
are indications that a careful study of the distribution
in China and Japan may prove of importance.

Moreover, the investigation of a smaller area may
also bring to light points of difficulty. In Ceylon, for
example, *P. polytes* is common up to several thousand
feet, while *P. hector* is rare at half the height to which
polytes ascends. Nevertheless the *H* form of female is
relatively just as abundant up-country where *hector*
is rarely found as it is low down where *hector* is plentiful[1]. On the other hand, *P. aristolochiae* may be exceedingly abundant at altitudes where *hector* is scarce. Yet
the *A* form of *polytes* is no more relatively abundant

[1] *Spolia Zeylanica*, 1910.

here than elsewhere on the island. All over Ceylon, in fact, the relative proportions of the three forms of female appear to be the same, quite irrespective of the abundance or scarcity of either of the models. As, however, we shall have to return to this point later, we may leave it for the moment to consider other features of this case of *P. polytes*.

In collections of insects from India or Ceylon it is not unusual to find specimens of the *A* form of female of *polytes* placed with *P. aristolochiae*, and the *H* form with *P. hector*. When the insects are old and faded and pinned out on cork the mistake is a very natural one. But after all the enemies of *polytes* do not hunt it in corked cabinets, and any estimation of resemblance to be of use to us must be based upon the living insects. Are the resemblances of the mimics to the models when alive so close that they might be expected to deceive such enemies[1] as prey upon them and have no difficulty in distinguishing the male form of *polytes* from *P. aristolochiae* or *P. hector* ?

To answer for a bird is a hazardous undertaking. We know so little of the bird's perceptive faculties whether of taste or sight. But on general grounds, from the specialization of their visual apparatus, it is probable that the sense of sight is keen, though whether the colour sense is the same as our own is doubtful[2]. On the other hand, the olfactory apparatus

---

[1] We shall take it for the present that, from the point of view of mimicry, birds are the main enemies of butterflies (cf. Chap. IX).

[2] See later, p. 119.

is relatively poorly developed in birds, and from this
we can only argue that the senses of smell and taste
are not especially acute.   Really we can do little more
than to describe how these mimetic resemblances
appear to our own senses, and to infer that they do not
appear very different to the bird.   If there is any
difference in keenness of perception we shall probably
not be far wrong in presuming that the advantage
rests with the bird.   After all if there is any truth in
the theory of mimicry the bird has to depend largely
upon its keenness of sight in making its living, at
any rate if that living is to be a palatable one.   If
natural selection can bring about these close resemblances among butterflies it must certainly be supposed
to be capable of bringing the bird's powers of vision to
a high pitch of excellence.

Returning now to the case of *P. polytes*, there is
not the least doubt that to the ordinary man accustomed
to use his eyes the *A* form of female is easily distinguish-
able from *P. aristolochiae*, as also is the *H* form from
P. hector.   The two models have a feature in common
in which they both differ from their respective mimics.
In both of them the body and head are largely of a
brilliant scarlet, whereas neither of the mimics has a
touch of red on the body.   In the living insect when
the body is swelled by its natural juices the effect is
very striking[1].   It gives at once a "dangerous" look

[1] The specimens figured on Pl. V were dried in papers when taken.
The body is consequently much compressed and the characteristic
scarlet of *P. hector* and *P. aristolochiae* is largely hidden.

to the insect when settled, even at a distance of several yards, and this although one may be perfectly familiar with its harmless nature. The mimics on the other hand with their sombre-coloured bodies never look otherwise than the inoffensive creatures that they are. The "dangerous" look due to the brilliant scarlet of the body and head of *hector* and *aristolochiae* is reinforced by the quality of the red on the markings of the wings. In both models it is a strong clamorous red suggestive of a powerful aniline dye, whereas such red as occurs in the mimics is a softer and totally distinct colour. The difference in quality is even more marked on the under than on the upper surface (Pl. V, figs. 3 *a*—6 *a*), and the net result is that when settled, with wings either expanded or closed, there is no possibility of an ordinarily observant man mistaking mimic for model in either case, even at a distance of several yards.

It may, however, be argued that it is not when at rest but during flight that the mimetic resemblance protects the mimic from attack. Actually this can hardly be true, for the mode of flight constitutes one of the most striking differences between model and mimic. *P. hector* and *P. aristolochiae* fly much in the same way. They give one the impression of flying mainly with their fore wings, which vibrate rapidly, so that the course of the insect, though not swift, is on the whole sustained and even. The flight of all the different forms of *polytes* is similar and quite distinct from that of the models. It is a strong but rather

heavy and lumbering up-and-down flight. One gets
the impression that all the wing surface is being used
instead of principally the fore wings as appears in
P. hector and *P. aristolochiae*. The difference is difficult
to put into words, but owing to these peculiarities of
flight the eye has no difficulty in distinguishing between
model and mimic even at a distance of 40 to 50 yards.
Moreover, colour need not enter into the matter at all.
It is even easier to distinguish model from mimic when
flying against a bright background, as for instance when
the insect is between the observer and a sunlit sky,
than it is to do so by reflected light. I have myself
spent many days in doing little else but chasing *polytes*
at Trincomalee where it was flying in company with
P. hector, but I was never once lured into chasing the
model in mistake for the mimic. My experience was
that whether at rest or flying the species are perfectly
distinct, and I find it difficult to imagine that a bird
whose living depended in part upon its ability to discriminate between the different forms would be likely
to be misled. Certainly it would not be if its powers
of discrimination were equal to those of an ordinary
civilised man. If the bird were unable to distinguish
between say the *A* form of female and *P. aristolochiae*
I think that it would be still less likely to distinguish
between the same *A* form and the male or the *M* form
of female. For my experience was that at a little
distance one could easily confuse the *A* form of
polytes with the male. Except when one was quite
close the red on the *A* form was apt to be lost, the

6—2

white markings on the hind wing were readily confused with those of the male, and one had to depend entirely on the lighter fore wing.   Unless the bird were keener sighted than the man the *A* form would be more likely to be taken in mistake for its unprotected relative than avoided for its resemblance to the presumably unpalatable model.   On the other hand, if the bird were sufficiently keen sighted never to confuse the *A* female with the male form its sight would be too keen to be imposed upon by such resemblance as exists between the *A* female and *P. aristolochiae.*

These, however, are not the only criticisms of the theory of mimicry which the study of this species forces upon us.   *Papilio polytes* is one of the few mimetic species that has been bred, and in no other case of polymorphism is the relation between the different forms so clearly understood.   For this result we are indebted mainly to the careful experiments of Mr J. C. F. Fryer, who recently devoted the best part of two years to breeding the different forms of this butterfly in Ceylon[1].   Fryer came to the conclusion that an explanation of this curious case is possible on ordinary Mendelian lines.   At first sight the breeding results appear complicated, for any one of the three forms of female can behave in several different ways.   For the sake of simplicity we may for the moment class together the *A* and *H* females as the mimetic females, the nonmimetic being represented by the *M* or male-like females.

---

[1] *Philosophical Transactions of the Royal Society,* vol. 204, 1913.

The different kinds of families which each of the three females can produce may be tabulated as follows:—

(α) The *M* form may give either:—

(1) *M* only.

(2) *M* and mimetics in about equal numbers.

(3) Mimetics only.

(β) The *A* form may give either:—

(1) *M* and mimetics in about equal numbers.

(2) *M* and mimetics in the ratio of about 1 : 3.

(3) Mimetics only.

(γ) The *H* form may give either:—

(1) *M* and mimetics in about equal numbers.

(2) *M* and mimetics in the ratio of about 1 : 3.

(3) Mimetics only.

The males are in all cases alike to look at but it must nevertheless be supposed that they differ in their transmitting powers. In fact the evidence all points to there being three different kinds of male corresponding to the three different kinds of female. But they cannot shew any difference outwardly because there is always present in the male a factor which inhibits the production of the mimetic pattern even though the factor for that pattern be present.

Returning now to the records of the females it will be noticed that although the *M* form may breed true the mimetics never give the *M* form alone. Where they give the *M* form among their progeny they produce mimetics and non-mimetics either in the ratio 1 : 1 or of 3 : 1. This at once suggests that the non-mimetic is recessive to the mimetic forms—that the mimetics contain a factor which does not occur in the non-mimetics. If this factor, which may be called X, be added to the constitution of a non-mimetic female it turns it into a mimetic. If X be added to a male such an individual, though incapable of itself exhibiting the mimetic pattern owing to the inhibitory factor always present in that sex, becomes capable of transmitting the mimetic factor to its offspring. Expressed in the usual Mendelian way the formulae for these different butterflies are as follows:—

$$\begin{array}{llll} M\,♀ & = & ii\,xx & Ii\,xx = ♂(1) \\ \text{Mimetic} \} & = \{ & ii\,XX & Ii\,XX = ♂(2) \\ ♀♀ & & \text{or } ii\,Xx & Ii\,Xx = ♂(3) \end{array}$$

where X stands for the mimetic factor and I for the factor which inhibits the action of X. All males are heterozygous for I, but during the segregation of characters at some stage in the formation of the families only the male-producing sperms come to contain the factor I. It is lacking in all the female-producing sperms formed by the male.

♂ (1) does not contain the factor for the mimetic condition and gives only daughters of the M form when mated with an M♀. ♂ (2) on the other hand is homozygous for the factor X, and consequently all of his germ cells contain it. This is the male that gives nothing but mimetic daughters with whatever form of female he is bred. ♂ (3) is heterozygous for X; that is to say, one half of his germ cells contain it, the other half not. With the M♀ he must give equal numbers of offspring with and without X, *i.e.* half of his daughters will be mimetic and the other half non-mimetic. With a heterozygous mimetic female ($iiXx$), which is also producing germ cells with and without X in equal numbers, he may be expected to give the usual result, viz. dominants and recessives in the ratio 3 : 1; or in other words mimetic and non-mimetic females in the ratio 3 : 1.

One of Fryer's experiments may be given here in illustration of the nature of the evidence upon which the above hypothesis depends.

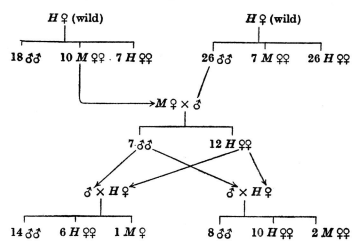

Families were reared from the two wild H females of whom nothing was known either as to ancestry or husband. The first family contained 10 M and 7 H females. Hence the original wild mother was probably $iiXx$ and had mated with a male of the constitution

Iixx. The family from the second wild *H* female contained 26 *H* and 7 *M* females; *i.e.* the ratio in which these two forms appeared was not far from 3 : 1. Hence the wild female was probably *iiXx* and her husband *IiXx*. If this were so some of the 26 ♂♂ should receive the *X* factor from both parents and consequently be *IiXX* in constitution. This was almost certainly so in the case of the single male in this brood tested by mating with an *M* female from the other brood. All of his 12 daughters were of the *H* form, as should have been the case had his constitution been *IiXX*. Supposing this to be so, all his offspring, of both sexes, must be heterozygous for *X*. Consequently any pair mated together should give both *H* and *M* females in the ratio of three of the former to one of the latter. In Mr Fryer's experiment two males and two females chosen at random were mated together. In the one case six *H* and one *M* female were produced, in the other ten *H* and two *M* females. As was expected both classes of female appeared, and the looked-for ratio of three *H* to one *M* was, in view of the smallness of the numbers, not departed from widely in either instance.

In the experiments selected as an illustration, the mimetic females happen to be all of the *H* form. In other experiments, however, both the *H* form and the *A* form occurred. As the result of his experiments Mr Fryer came to the conclusion that here again the difference is one of a single hereditary factor. All mimetic females contain the *X* factor, but the *H* females contain in addition a factor which we may call Y. The function of the Y factor is to carry the change made by the X factor a step further, and to turn the A form of female into the H form. Y is a modifier of X, but unless X is present Y can produce no effect. All the different individuals which are to be found among *P. polytes* in Ceylon may be represented as follows:—

♂♂	M ♀♀	A ♀♀	H ♀♀
Ii xx YY	*ii xx YY*	—	—
Ii xx Yy	*ii xx Yy*	—	—
Ii xx yy	*ii xx yy*	—	—
Ii Xx YY	—	—	*ii Xx YY*
Ii Xx Yy	—	—	*ii Xx Yy*
Ii Xx yy	—	*ii Xx yy*	—
Ii XX YY	—	—	*ii XX YY*
Ii XX Yy	—	—	*ii XX Yy*
Ii XX yy	—	*ii XX yy*	—

In this way is offered a simple explanation in terms of three Mendelian factors which serves at once to explain the various results of the breeding experiments, and the fact that intermediates between the different forms of female are not found.

The only other experiments comparable with these on *P. polytes* are some made by Jacobsen on *Papilio memnon* in Java[1]. Here again there are three forms of female, one of which, *laomedon*, is something like the male, while the other two, *agenor* and *achates*, are quite distinct. Of these three *achates*, unlike the male and the other two females, is tailed, and resembles

[1] *Tijdschr. voor Entomologie*, vol. 53, 1909. A more accessible account is given by de Meijere, *Zeit. f. indukt. Abstamm. u. Vererbungslehre*, vol. 3, 1910.

the species *Papilio coon* which belongs to the same presumably distasteful group as *P. aristolochiae*. These experiments of Jacobsen's are not so complete as the series on *P. polytes*, but Professor de Meijere and Mr Fryer have both pointed out that they are capable of being interpreted on the same simple lines.

Another instance of experimental breeding involving polymorphism and mimicry in the female sex is that of the African *Papilio dardanus*, but the case is here complicated by the greater number of female forms (cf. pp. 30–33). The data, too, are far more scanty than in the other two cases, but so far as they go there is nothing to preclude an explanation being eventually arrived at on similar lines[1].

And now we may consider briefly the bearing of these experiments on the theory of mimicry. Throughout the work no individuals intermediate between the three well-marked forms of *polytes* were met with. There is no difference in appearance between the heterozygous and the homozygous mimetic insects, whether they belong to the *A* or to the *H* form. The factor *X*, whether inherited from both parents, or from one only, produces its full effect, and the same is also true of the action of the factor *Y*. Now the most generally accepted hypothesis as to the formation of these mimetic resemblances supposes that they have been brought about through the gradual operation of natural selection accumulating slight variations.

[1] For further information see Poulton, *Trans. Ent. Soc. Lond.* 1909, and various notes in *Proc. Ent. Soc. Lond.* subsequent to this date.

Professor Poulton, for example, a prominent exponent of this school, considers that the *A* form of female was first evolved gradually from the *M* form, and later on the *H* form came by degrees from the *A* form. If this be true we ought, by mingling the *M* germ plasm with the *H* germ plasm and by subsequently breeding from the insects produced, to get back our series of hypothetical intermediates, or at any rate some of them. We ought as it were to reverse the process by which the evolution of the different forms has taken place. But as is shewn by the experiment of Mr Fryer, which was quoted above, nothing of the sort happens.

From experiments with cultivated plants such as primulas and sweet peas, we have learnt that this discontinuous form of inheritance which occurs in *P. polytes* is the regular thing. Moreover, we have plenty of historical evidence that the new character which behaves in this way is one that has arisen suddenly without the formation of intermediate steps. The dwarf "Cupid" form of sweet pea, for instance, behaves in heredity towards the normal form as though the difference between them were a difference of a single factor. It is quite certain that the "Cupid" arose as a sudden sport from the normal without the intervention of anything in the way of intermediates. And there is every reason to suppose that the same is true for plenty of other characters involving colour and pattern as well as structure, both in the sweet pea, the primula, and other species. Since the forms of *polytes* female behave in breeding like the various forms of sweet pea and primula there is every reason to suppose that they *arose* in the same way, that is to say, as sudden sports or mutations and not by the gradual accumulation of slight differences.

But if we take this view, which is certainly most consonant with the evidence before us, we must assign to natural selection a different rôle from that which is generally ascribed to it. We cannot suppose that natural selection has played any part in the *formation* of a mimetic likeness. The likeness turned up suddenly as a sport quite independently of natural selection. But although natural selection may have had nothing to do with its production, it may nevertheless have come into play in connection with the *conservation* of the new form. If the new form possesses some advantage over the pre-existing one from which it sprang, is it not conceivable that natural selection will come into operation to render it the predominant form? To this question we shall try to find an answer in the next chapter.

CHAPTER VIII

THE CASE OF *PAPILIO POLYTES* (*cont.*)

IT was suggested in the last chapter that if a new variation arose as a sport—as a sudden hereditary variation—and if that variation were, through resemblance to a different and unpalatable species, to be more immune to the attacks of enemies than the normal form, it was conceivable that the newer mimetic sport would become established, and in time perhaps come to be the only form of the species. We may suppose, for example, that the *A* female of *P. polytes* arose suddenly, and that owing to its likeness to the presumably distasteful *P. aristolochiae* it became rapidly more numerous until in some localities it is the commonest or even the only form. However, before discussing the establishing of a mimetic form in this manner we must first deal with certain general results which may be expected to follow on a process of selection applied to members of a population presenting variations which are inherited on ordinary Mendelian lines.

Let us suppose that we are dealing with the inheritance of a character which depends upon the presence of the genetic factor X; and let us also suppose that the heterozygous form (Xx) is indistinguishable from the homozygous form (XX) in appearance. In other words the character dependent upon X exhibits complete dominance. With regard to X then all the members of our population must belong to one or other of three classes. They may be homozygous (XX) for X, having received it from both parents, or they may be heterozygous (Xx) because they have received it from only one parent, or they may be devoid of X, *i.e.* pure recessives (xx). An interesting question arises as to the conditions under which a population containing these three kinds of individuals remains stable. By stability is meant that with the three kinds mating freely among themselves and being all equally fertile, there is no tendency for the relative proportions of the three classes to be disturbed from generation to generation. The question was looked into some years ago by G. H. Hardy, who shewed that if the mixed population consist of p XX individuals, $2q$ Xx individuals and r xx individuals, the population will be in stable equilibrium with regard to the relative proportions of these three classes so long as the equation $pr = q^2$ is satisfied[1].

Now let us suppose that in place of equality of conditions selection is exercised in favour of those individuals which exhibit the dominant character. It has been shewn by Mr Norton that even if the selection exercised were slight the result in the end would be that the recessive form would entirely disappear. The total time required for bringing this about would

[1] *Science,* July, 1908.

depend upon two things, (1) the proportion of dominants existing in the population before the process of selection began, and (2) the intensity of the selection process itself. Suppose, for example, that we started with a population consisting of pure dominants, heterozygotes, and recessives in the ratio 1 : 4 : 4. Since these figures satisfy the equation $pr = q^2$, such a population mating at random within itself is in a state of stable equilibrium. Now let us suppose that the dominant form (including of course the heterozygotes) is endowed with a selection advantage over the recessives of 10 %, or in other words that the relative proportion of the recessives who survive to breed is only 90 % of the proportion of dominants that survive[1]. It is clear that the proportion of dominants must gradually increase and that of the recessives diminish.

At what rate will this change in the population take place? Mr Norton has worked this out (see App. I) and has shewn that at the end of 12 generations the proportions of pure dominants, heterozygotes, and recessives will be 1 : 2 : 1. The population will have reached another position of equilibrium, but the proportion of recessives from being four-ninths of the

[1] If for example there were 5000 dominants and 4000 recessives, and if only half of the population survives to mate, then we should be left with 2500 dominants and 2000 recessives as parents of the next generation. But if there were also a 10 % selective disadvantage working against the recessives, their numbers would be further reduced from 2000 to 1800 and the proportion of dominants to recessives would be changed from 5 : 4 to 25 : 18.

total is now reduced to one-quarter. After 18 more generations the proportions 4 : 4 : 1 are reached, the recessives being only one-ninth of the total; after 40 further generations of the process they become reduced to one-fortieth. In other words a selective advantage of 10% operating against the recessives will reduce their numbers in 70 generations from nearly one-half of the population to less than one-fortieth.

With a less stringent selective rate the number of generations elapsing before this result is brought about will be larger. If, for example, the selective rate is diminished from 10% to 1% the number of generations necessary for bringing about the same change is nearly 700 instead of 70—roughly ten times as great. Even so, and one can hardly speak of a 1% selective rate as a stringent one, it is remarkable in how brief a space of time a form which is discriminated against, even lightly, is bound to disappear. Evolution, in so far as it consists of the supplanting of one form by another, may be a very much more rapid process than has hitherto been suspected, for natural selection, if appreciable, must be held to operate with extraordinary swiftness where it is given established variations with which to work.

We may now consider the bearing of these theoretical deductions upon the case of *Papilio polytes* in Ceylon. Here is a case of a population living and breeding together under the same conditions, a population in which there are three classes depending upon the presence or absence of two factors, X and Y, exhibiting ordinary Mendelian inheritance. For the present we may consider one of these factors, X, which involves the proportion of mimetic to non-mimetic forms. It is generally agreed among observers who have studied this species that of the three forms of female the M form is distinctly the most common, while of the other two the H form is rather more numerous than the A form. The two dominant mimetic forms taken together, however, are rather more numerous than the recessive M form. The most recent observer who studied this question, Mr Fryer, captured 155 specimens in the wild state as larvae. When reared 66 turned out to be males, while of the females there were 49 of the two mimetic forms and 40 of the M form, the ratio of dominants to recessives being closely $5 : 4$[1]. Now as has already been pointed out the ratio $5 : 4$ of dominants and recessives is characteristic of a population exhibiting simple Mendelian inheritance when in a state of stable equilibrium. The natural deduction from Mr Fryer's figures is that with regard to the factor that differentiates the mimetic forms from the non-mimetic, the *polytes* population is, for the moment at any rate, in a position of stable equilibrium. This may mean one of two things. Either the population is definitely in a state of equilibrium which has lasted for a period of time in the past

---

[1] As these larvae were for the most part found simply over a considerable time it follows that they are the offspring of different females and represent the relative proportions of the three forms in the general population.

and may be expected to endure for a further period in the future, or else the population is in a condition of gradual change as regards the numerical proportion of mimetics and non-mimetics, progressing towards the elimination of the one or the other, the present state of equilibrium being merely transitory and accidental. In this connection a few scraps of historical evidence are of interest. Of the various forms of *P. polytes* the *A* form of female was the first to be described in 1758, and not long after (1776) the *H* form was registered as a species under the name of *Papilio Eques Trojanus romulus*. Later on the female resembling the male found its way into the literature as *Papilio pammon*. From the fact that the mimetic forms were known before the non-mimetic, it is unlikely that they can have been scarce a century and a half ago. As *P. polytes* certainly produces at least four broods a year in Ceylon this period of time represents something like 600 generations in the life of the species, and we have already seen that even if the mimetic forms have but a 1 % advantage over the non-mimetic the proportion of the latter would decrease from nearly equality down to but 1 in 40 in about 700 generations. Actually for *P. polytes* the decrease would not be so marked because the male is non-mimetic. Owing to this peculiar feature the rapidity of change in the proportion of the different forms is reduced to about one-half of what it would be if the males were also mimetic. Nevertheless the change from nearly equality to about one non-mimetic in 40 would have taken place

during the time *P. polytes* has been known if a 2%
selection advantage had operated during that period
in favour of the mimetic. If there has been any
appreciable selection going on during that time mimetics must have been far rarer when the species was
first discovered, but the fact that both the mimetic
forms made their way into collections before the
non-mimetic tells distinctly against this supposition.
Nor is there any reason to suppose that the non-mimetic
form has been dwindling in numbers relatively to the
mimetics during the last half century. Moore[1] in 1880
records an earlier observation of Wade's that "These
three butterflies are very common, especially those of
the first form; the second being perhaps least so."
The first form alluded to is the *M* form, and the
second is the *A* form, so that at the time Wade wrote
the relative proportions of these three forms must
have been very much what they are to-day. Even
during half a century and with such a relatively weak
selection rate as 2% in favour of the mimetics, the
proportion of non-mimetics should drop from about
4 : 5 down to about 1 : 5. Therefore we must either
infer that in respect of mimetic resemblances natural
selection does not exist for *P. polytes* in Ceylon, or else
we must suppose its force to be so slight that in half a
century certainly, and perhaps in a century and a
half, it can produce no effect appreciable to the necessarily rough method of estimation employed.

[1] *The Lepidoptera of Ceylon*, 1880.

It may, however, be argued that even an exceedingly low selection rate is able to bring about the elimination of one or other type provided that it acts for a sufficiently long time.   This is perfectly true.   A selective rate of ·001 % would reduce the proportion of recessives to dominants from 4 : 5 down to 1 : 40 in the course of about 1,400,000 generations where the mimetic resemblance is already established.   Such a form of selection entails the death of but one additional non-mimetic in 100,000 in each generation.   If, however, the mimetic resemblance is not fully established and the mimic bears only what supporters of the mimicry theory term a "rough" resemblance to the model, it is clear that it will have far less chance of being mistaken for the model.   Its advantage as compared with the non-mimetic form will be very much less.   Even supposing that the slight variations concerned are inherited, an intensity of selection which would produce a certain change in 1,400,000 generations where a mimetic resemblance is already established must be supposed to take an enormously greater time where an approach to a model has to take place from a "rough" resemblance.

From the data as to the relative proportions of the polymorphic females of *P. polytes* during the past and at present, and from the behaviour of their different forms in breeding, the following conclusions only can be drawn.   Either natural selection, from the point of view of mimicry, is non-existent for this species in Ceylon, or else it is so slight as to be unable in half a century to produce an appreciable diminution in the proportion of non-mimetic females. For even if the mimetic resemblance brings about but the survival of one additional protected form in 100 as compared with the unprotected, this means a marked diminution in the proportion of *M* females in 50 years—a diminution such as there are no grounds for supposing to have taken place.

It has been argued that in populations exhibiting Mendelian heredity even a relatively low selection rate must bring about a rapid change in the constitution of a mixed population. Have we any grounds for supposing that populations of this sort can undergo such rapid changes? In cases where mimetic resemblances are involved we have no examples of the sort. But some interesting evidence as to the rate at which a population may change is to be gathered from the study of melanism in certain moths. It is well known that in some parts of England the common peppered moth, *Amphidasys betularia* has been almost entirely supplanted by the darker melanic form *doubledayaria*. It first made its appearance near Manchester in 1850, and from that centre has been gradually spreading over northern England, the Midlands, and the south-eastern counties. At Huddersfield, for instance, fifty years ago only the type form *betularia* existed; to-day there is nothing but *doubledayaria*. In Lancashire and Cheshire the type is now rare. On the continent, too, there is the same story to be told. The melanic form first appeared in Rhenish

Prussia in 1888; to-day it is much more abundant than the older type. There, too, it is spreading eastwards and southwards to Thuringia, to Saxony, to Silesia. What advantage this new dark form has over the older one we do not know[1]. Some advantage, however, it must have, otherwise it could hardly supplant *betularia* in the way that it is doing. From our present standpoint two things are of interest in the case of the peppered moth—the rapidity with which the change in the nature of the population has taken place, and the fact that the two forms exhibit Mendelian heredity, *doubledayaria* being dominant and *betularia* recessive[2]. Moreover, mixed broods have been reared from wild females of both sorts, and so far as is known the two forms breed freely together where they co-exist. This case of the peppered moth shews how swiftly a change may come over a species[3]. It is not at all improbable that the establishing of a new variety at the expense of an older one in a relatively short space of time is continually going on, especially in tropical lands where

[1] From the experience of breeders it would appear that the melanic form is somewhat hardier, at any rate in captivity.

[2] Intermediates may also occur in some strains (cf. Bowater, *Journal of Genetics*, vol. 3, no. 4, 1914).

[3] An interesting case of a similar nature has recently been published by Hasebroek (*Die Umschau*, 1913, p. 1020). A melanic form of the moth, *Cymatophora or*, suddenly appeared near Hamburg in 1904. This new form, to which the name *albingensis* was given, rapidly became the predominant one. In 1911–1912 over 90 % of the moths reared from caterpillars taken in the open were of the *albingensis* form; nor were any intermediates found between it and the typical form. Some experiments were also made which shew that the *albingensis* form behaves as a dominant to the original type form.

the conditions appear to be more favourable to exuberance of variation and where generations succeed one another in more rapid succession.   At present, however, we are without data.   A form reported by an old collector as common is now rare; a variety once regarded as a great prize is now easily to be found.   Such to-day is the sort of information available.   For the solution of our problem it is, of course, useless.   The development of Mendelian studies has given us a method, rough perhaps but the best yet found, of testing for the presence, and of measuring the intensity, of natural selection.   Much could be learned if some common form were chosen for investigation in which, as in *P. polytes*, there are both mimetic and non-mimetic forms.   Large numbers should be caught at stated intervals, large enough to give trustworthy data as to the proportions of the different forms, mimetic or non-mimetic, that occurred in the population.   Such a census of a polymorphic species, if done thoroughly, and done over a series of years at regular intervals, might be expected to give us the necessary data for deciding whether the relative proportion of the different forms was changing—whether there were definite grounds for supposing natural selection to be at work, and if so what was the rate at which it brought the change about.

CHAPTER IX

THE ENEMIES OF BUTTERFLIES

THE theory of mimicry demands that butterflies should have enemies, and further that those enemies should exercise a certain discrimination in their attacks. They must be sufficiently observant to notice the difference between the mimetic and the non-mimetic form; they must be sufficiently unobservant to confuse the mimetic form with the unpalatable model. And, of course, they must have enough sense of taste to dislike the unpalatable and to appreciate the palatable varieties. What these enemies are and whether they can be supposed to play the part required of them we may now go on to consider.

Butterflies are destroyed in the imago state principally by three groups of enemies—predaceous insects, lizards, and birds. It is known that monkeys also devour butterflies to some extent, but such damage as they inflict is almost certainly small in comparison with that brought about by the three groups already mentioned. In view of the very different nature of these groups it will be convenient to consider them separately.

I. *Predaceous Insects.* Butterflies are known to be preyed upon by other insects of different orders, and a considerable number of observations have recently been gathered together from various sources and put on record by Professor Poulton[1]. These observations shew that butterflies may be devoured by mantids, dragon-flies, and blood-sucking flies of the families Empiidae and Asilidae. For mantids the records are scanty, but they have been observed to kill presumably distasteful forms as often as those which are considered palatable. An interesting set of experiments was made by G. A. K. Marshall on captive mantids in Africa[2]. Of the eleven individuals representing several species with which he experimented, some ate every butterfly offered, including the distasteful Danaines and Acraeines. Others, however, shewed some distaste of the Acraeines and would not devour them so freely as butterflies of other species. There are no grounds, however, for supposing that the mantids had any appreciation of the warning coloration of the Acraeines. Whether completely eaten or not the Acraeines were apparently sufficiently damaged to prevent their taking any further part in the propagation of their species. Warning coloration is not of much service to its possessor who has to be tasted and partially eaten before being eventually rejected. Even if some mantids shew distaste of certain unpalatable butterflies, that distaste is probably seldom

[1] *Trans. Ent. Soc. Lond.* 1907.
[2] *Trans. Ent. Soc. Lond.* 1902.

exercised with a gentleness sufficient to ensure that the butterfly reaps the reward of its disagreeable nature. And unless, of course, the butterfly is allowed to do so the enemy can play no part in the production or maintenance of a mimetic resemblance.

What is true for mantids is probably also true for the other groups of predaceous insects. Dragon-flies and wasps have been recorded as attacking the distasteful as well as butterflies of unprotected groups. Among the most serious enemies of butterflies must probably be reckoned the blood-sucking Asilids. These powerful and ferocious flies seize butterflies on the wing with their strong claws and plunge their proboscis into the thorax. Apparently they inject some swift poison, for the butterfly is instantly paralysed, nor is there any sign of struggle. The Asilid flies off with its victim, sucking the juices as it goes. There can be no doubt in the mind of any one who has watched these creatures hawking butterflies that their natural gifts are such as to enable them to exercise discrimination in their food. Most insect life is at their mercy but they appear to exercise no choice, seizing and devouring the first flying thing that comes within easy reach. Certainly as regards butterflies palatability or the reverse makes no difference, and they are known to feed indiscriminately both upon the evil-flavoured and upon the good. Taking it all together the evidence is such that we cannot suppose predaceous insects to pay any attention to warning colours, and, therefore, we cannot regard them as playing any part in connection with mimetic resemblance.

II. *Lizards.* In those parts of the world where
lizards of larger size are abundant there is plenty of
evidence that certain species are very destructive to
butterfly life. As might be expected this is especially
true of forms which are either arboreal or semi-arboreal
in habit. Among the reptiles of Ceylon, for example,
are several species of the genus *Calotes*, of which two,
C. ophiomachus and *C. versicolor*, are particularly abundant. In appearance and habits they are not unlike
chameleons though far more active in their movements.
Like chameleons, too, they are able to change colour,
and the fact that they can assume a brilliant scarlet
hue about the head and neck has probably led to their
popular name of "blood-suckers." It is not impossible
that the assumption of this scarlet coloration may
serve as a lure to bring insects within range. These
lizards have often been observed to seize and devour
butterflies. Moreover, it is a common thing to find
butterflies with a large semi-circular patch bitten out
of the hind wings, and there is little doubt but
that such injuries have been inflicted by lizards.
There is, however, no evidence to suggest that they
exercise any discrimination in their choice of the
butterflies which they attack. This is borne out by
their behaviour towards various species offered to
them, both when at liberty and when caged. In an
ingenious series of experiments Col. Manders brought
various butterflies within reach of a *Calotes* by the
help of a fishing-rod and a long line of fine silk, by this
means simulating natural conditions as far as possible.

He found that the lizards ate the so-called distasteful forms such as *Danais chrysippus, Euploea core, Acraea violae,* and *Papilio hector,* as readily as the presumably more palatable forms[1]. In captivity, too, they will take any butterfly as readily as another. Experiments by Finn[2] and by the writer[3] proved that they ate Danaids, Euploeas, and *Papilio aristolochiae* without any hesitation so long as the insects were alive and moving. When, too, a mixture of different species, some with and some without warning coloration, was given to them all were eaten, nor was there any discrimination evidenced in the order in which they were taken. The lizard simply took the first that came within reach and went on until the whole lot was devoured, wings and all.

Some experiments by Miss Pritchett on the American lizard *Sceleporus floridanus* point to the same conclusion[4]. She found that it took without hesitation any butterfly offered to it including the presumably distasteful models *Danais archippus* and *Papilio philenor* (cf. pp. 45 and 49). On the other hand, another species of lizard with which Miss Pritchett experimented, *Gerrhonotus infernalis,* refused all the butterflies offered to it, though it fed freely on Orthopterous insects as well as on spiders and scorpions.

It seems clear from these various observations and

[1] *Proc. Zool. Soc.* 1911.
[2] *Journ. Roy. Asiat. Soc. Bengal,* vol. 65, 1897.
[3] *Spolia Zeylanica,* 1910.
[4] *Biological Bulletin,* vol. 5, 1903.

experiments that certain lizards devour butterflies freely, but that they do not exercise any discrimination in the species which they attack. All are caught and devoured indiscriminately, so that in spite of the fact that such lizards are among the most serious enemies of butterflies we cannot suppose them to play any part in establishing a mimetic resemblance.

III. *Birds.* The relations which exist between butterflies and their bird enemies have for many years been the subject of keen discussion. It is generally recognised that if mimetic resemblances become established through the agency of discriminating enemies those enemies must be birds. Hence those interested in the question of mimicry have for some years past turned their attention to birds more than to the other enemies of butterflies. That many birds systematically feed on butterflies is a fact that does not admit of doubt. It is true that, as Mr Marshall points out in the valuable paper in which he has summarised the evidence[1], observations of birds eating butterflies are relatively scanty. Though, as he points out, this is equally true for other groups of insects besides butterflies, bird attacks on butterflies, owing to the conspicuous nature of the victim, are much more likely to attract attention than attacks on other groups. We are still without much information as to the extent to which birds destroy butterflies and as to whether they shew any decided preference for certain species over others. A careful examination of the contents of the

[1] *Trans. Ent. Soc. Lond.* 1909.

stomachs of large numbers of insectivorous birds in a tropical area would go some way towards deciding the matter, but at present such information is lacking. We have to rely upon the existing observations of birds attacking butterflies in the wild state, and upon certain feeding experiments made with captive birds.

Observations on birds attacking butterflies where mimetic forms occur have been made almost entirely in certain parts of Africa, in India, and in Ceylon. For Africa, Marshall has collected some forty-six observations of which almost half are concerned with Pierines. The remainder include four instances of attacks on species of *Acraea*, a genus which on the mimicry theory must be regarded as among the most unpalatable of butterflies.

The records from the Indo-Malayan region (principally India and Ceylon) are somewhat more numerous and here again more than one-third of them refer to Pierines. Among the others are records of the distasteful forms *Euploea core, E. rafflesii, Acraea violae,* and *Papilio hector* being taken and devoured.

There is one interesting record which seems to suggest that Swinhoe's Bee-Eater (*Melittophagus swinhoei*) may exercise that discrimination in the butterflies it attacks which is demanded on the mimicry theory. Lt.-Col. Bingham on one occasion in Burma noticed this species hawking butterflies. He records that they took *Papilio erithonius, P. sarpedon, Charaxes athamas, Cyrestis thyodamas,* and *Terias hecabe,* and probably also species of the genera *Prioneris, Hebomoia* (Pierines),

Junonia and *Precis* (Vanessids). And he goes on to say: "I also particularly noticed that the birds never went for a *Danais* or *Euploea*, or for *Papilio macareus* and *P. xenocles*, which are mimics of Danais, though two or three species of *Danais*, four or five of *Euploea*, and the two above-mentioned mimicking *Papilios* simply swarmed along the whole road[1]."

Marshall also quotes a case of attack by a green bee-eater on a *Danais* in which the butterfly was caught and subsequently rejected, after which it flew away. Little stress, however, can be laid upon this case in view of the more recent data brought together by Col. Manders and Mr Fryer. Discussing the attacks of birds on butterflies in Southern India and Ceylon, Col. Manders gives the following quotation[2] from a letter of Mr T. N. Hearsy, Indian Forest Service:

"Coimbatore, 6. 6. 10.... I have frequently seen the common green bee-eater (*Merops viridis*) and the king-crow (*Buchanga atra*) take butterflies on the wing, the butterflies being *Catopsilia pyranthe*, *C. florella*, *Terias hecabe* and *Papilio demoleus*. The bee-eater I have also seen taking *Danais chrysippus* and *Danais septentrionis*, and I remember to have been struck with their taste for those latter...."

Col. Manders also brings forward evidence for these Danaids and Euploeas being eaten by Drongos and by the paradise flycatcher. Still more recently an interesting contribution to the matter has been made by

---

[1] *Trans. Ent. Soc. Lond.* 1902.
[2] *Trans. Ent. Soc. Lond.* 1911.

Mr J. C. F. Fryer[1]. The Ashy Wood-swallow (*Artamus fuscus*) had been recorded on two occasions as having attacked *Euploea core*. Mr Fryer was fortunate in coming across this bird in the gardens at Peradeniya, near Kandy, at a time when *Euploea core* and *Danais septentrionis* were particularly abundant, and he watched a number of them systematically hawking these presumably unpalatable species. As he observes, "in Ceylon a resemblance to the genera *Danais* and *Euploea* is doubtfully of value; in fact in the neighbourhood of Wood-swallows it is a distinct danger." Fryer also noted that the mimetic forms of *P. polytes* were taken as well as the non-mimetic.

For tropical Central and South America, that other great region where mimetic forms are numerous, there are unfortunately hardly any records of butterflies attacked by birds. Bates stated that the Pierines were much persecuted by birds, and his statement is confirmed by Hahnel, but exact observations for this region are remarkably scanty. Belt observed a pair of birds bring butterflies and dragon-flies to their young, and noticed that they brought no Heliconii to the nest although these swarmed in the neighbourhood[2]. On the other hand, Mr W. Schaus[3], from an experience of many years spent in the forests of Central America, considers that the butterflies of this region are hardly, if ever, attacked by birds.

[1] *Proc. Zool. Soc.* 1913.
[2] *A Naturalist in Nicaragua*, 1874, p. 316.
[3] *Ier Congr. Internat. d'Entomologie*, Bruxelles, 1911.

For North America Marshall records over 80 cases of birds attacking butterflies. Among them is an interesting record of a bird seizing and rejecting a specimen of *Anosia plexippus* (= *Danais archippus*), one of the few Danaines found in this region.

It must be admitted that the data at present available with regard to the attacks of birds upon butterflies under natural conditions are too meagre to allow of our coming to definite conclusions on the points at issue. It is safe to say that a number of species of birds have been known to attack butterflies—that a few out of the number feed upon butterflies systematically—that some of the most persistent bird enemies devour the presumably protected forms as freely as the unprotected—but that in a few instances there is some reason for supposing that the bird discriminates. Beyond this it is unsafe to go at present.

In attempting to come to a decision as to the part played by birds in the destruction of butterflies an evident desideratum is a knowledge of the contents of the stomachs of freshly killed birds. Unfortunately few systematic observations of this nature exist. G. L. Bates[1], when collecting in the Southern Cameroons, noted the stomach contents of a considerable number of birds. The remains of beetles were recognised in 213 cases: Orthoptera in 177: ants in 57 (mostly in stomachs of birds of the genus *Dendromus*): other Hymenoptera in 8: coccids in 32: bugs in 19: white ants in 31: slugs and snails in 24: spiders in 85

---

[1] *Ibis*, 1911.

(mostly in Sunbirds): millipedes in 20; but in no single instance were the remains of butterflies found. More recently Bates' account has been criticized by Swynnerton[1] who comments on the difficulty of identifying butterfly remains as compared with those of beetles and grasshoppers. He states that the pellets ejected by captive birds after a meal of butterflies contain only fine debris which is very difficult to identify. Further, he found that of twenty small bird excreta collected in the forest no less than eighteen contained scales and small wing fragments of Lepidoptera.

Some attention has been paid to the relation between birds and butterflies in the United States, and under the auspices of the Department of Agriculture a large number of birds' stomachs have been investigated. Careful examination of some 40,000 stomachs of birds shot in their natural habitats resulted in the discovery of butterfly remains in but four. It cannot, therefore, be supposed that birds play much part in connection with such mimetic resemblances as are found in North America (cf. pp. 45–49). Nevertheless, it is known that on occasion large numbers of butterflies may be destroyed by birds. An interesting case is described by Bryant[2] of an outbreak in North California of *Eugonia californica*, a close relative of the tortoiseshell. The butterfly was so abundant as to be a plague, and five species of birds took advantage of its great abundance to prey largely upon it. From

[1] *Ibis*, 1912.
[2] *The Condor*, vol. 13, 1911, pp. 195–208.

his examination of the stomachs Bryant came to the conclusion that some 30 % of the food of these five species was composed of this butterfly. The stomachs of many other species were examined without ever encountering butterfly remains. Nor did field observations support the view that any species, other than the five specially noted, ever attacked these butterflies. The case is of interest in the present discussion as evidence that the identification of butterfly remains in the stomachs of birds is by no means so difficult as some observers suggest.

Besides this evidence derived from observations upon birds in the wild state some data have been accumulated from the experimental feeding of birds in captivity. Of such experiments the most extensive are those of Finn[1] in South India. He experimented with a number of species of insectivorous birds belonging to different groups. Of these he found that some, among which may be mentioned the King-crow, Starling, and Liothrix[2], objected to Danaines, *Papilio aristolochiae* and *Delias eucharis*, a presumably distasteful Pierine with bright red markings on the under surface of the hind wings (Pl. II, fig. 1). In some cases the bird refused these forms altogether, while in others they were eaten in the absence of more palatable forms. The different species of birds often differed in

---

[1] *Journ. Asiat. Soc. Bengal,* vol. 64, 1895, and vol. 66, 1897.

[2] Nevertheless a Liothrix is recorded as eating *Danais plexippus* and a *Euploea* even though two male specimens of the palatable *Elymnias undularis* were in the cage.

their behaviour towards these three "nauseous" forms. The Hornbill, for example, refused the Danaines and *P. aristolochiae* absolutely, but ate *Delias eucharis*. Some species again, notably the Bulbuls (*Molpastes*) and Mynahs, shewed little or no discrimination, but devoured the "protected" as readily as the "unprotected" forms. Finn also states that "*Papilio polytes* was not very generally popular with birds, but much preferred to its model, *P. aristolochiae*."

In many of Finn's experiments both model and mimic were given to the birds simultaneously so that they had a choice, and he says that "in several cases I saw the birds apparently deceived by mimicking butterflies. The Common Babbler was deceived by *Nepheronia hippia*[1] and Liothrix by *Hypolimnas misippus*. The latter bird saw through the disguise of the mimetic *Papilio polites*, which, however, was sufficient to deceive the Bhimraj and King-crow. I doubt if any bird was impressed by the mimetic appearance of the female *Elymnias undularis*" (cf. Pl. IV, fig. 5). Finn concluded from his experiments that on the whole they tended to support the theory of Bates and Wallace, though he admits that the unpalatable forms were commonly taken without the stimulus of actual hunger and generally without signs of dislike. Certainly it is as well to be cautious in drawing conclusions from experiments with captive birds. The King-crow, for instance, according to Finn shewed a marked dislike for Danaines in captivity; yet Manders records this

---

[1] A form closely resembling *P. ceylonica* figured on Pl. I, fig. 1.

species as feeding upon Danaines under natural conditions (cf. p. 111).

A few further experiments with the birds of this region were carried out by Manders[1] in Ceylon. The results are perhaps to be preferred to Finn's, as the birds were at liberty. Manders found that the Brown Shrike (*Lanius cristatus*) would take butterflies which were pinned to a paling. In this way it made off with the mimetic females of *Hypolimnas bolina* and *H. misippus*, as well as with *Danais chrysippus* and *Acraea violae* which were successively offered to it. Evidently this species had no repugnance to unpalatable forms. Manders also found that a young Mynah allowed complete liberty in a large garden would eat such forms as *Acraea violae* and *Papilio hector*. As the result of his experience Manders considers that the unpalatability of butterflies exhibiting warning coloration has been assumed on insufficient data, and he is further inclined to doubt whether future investigations will reveal any marked preference in those birds which are mainly instrumental in the destruction of butterflies.

A few experiments on feeding birds with South African butterflies are recorded by Marshall. A young Kestrel (*Cerchneis naumanni*) was fed from time to time with various species of butterflies. In most cases the butterflies offered were eaten even when they were species of *Acraea*. On the other hand *Danais chrysippus* was generally rejected after being partly

[1] *Proc. Zool. Soc. Lond.* 1911.

devoured. When first offered this unpalatable species
was taken readily and it was only after it had been
tasted that the bird rejected it. When offered on
several subsequent occasions it was partly eaten each
time, and the behaviour of the Kestrel did not suggest
that it associated a disagreeable flavour even with this
conspicuous pattern. Another young Kestrel (*Cerchneis rupicoloides*) was also used for experiment. At
first it would not take butterflies and at no time did
it shew any fondness for them. Indeed it is doubtful
from the way in which they seem to have shaped at
the insects whether either of these Kestrels had had any
experience of butterflies before the experiments began.

A Ground Hornbill with which Marshall also experimented ate various species, including *Acraea*, but,
after crushing it, refused the only *Danais chrysippus*
offered. It is hardly likely that this large omnivorous
bird operates as a selecting agent in cases of mimicry.

In an interesting paper published recently McAtee[1]
discusses the value of feeding experiments with animals
in captivity as a means of indicating their preference
for different articles of diet. After reviewing the
various evidence brought forward he concludes that
the food accepted or rejected by captive animals is
very little guide to its preferences under natural conditions. He points out that a bird in captivity not
infrequently rejects what is known to form a main
staple of its diet in nature, and that conversely it may
eagerly accept something which, in the wild state, it

[1] *Proc. Acad. Nat. Sci. Philadelphia*, 1912.

would have no opportunity of obtaining. Great caution must, therefore, be exercised in the interpretation of feeding experiments made with birds in captivity.

It appears to be generally assumed that colour perception in birds is similar to what it is among human beings, but some experiments made by Hess[1] render it very doubtful whether this is really the case. In one of these experiments a row of cooked white grains of rice was illuminated by the whole series of spectral colours from violet to deep red. Hens which had been previously kept in the dark so that their eyes were adapted to light of low intensity were then allowed to feed on the spectral rice. The grains illuminated by green, yellow, and red were quickly taken, but the very dark red, the violet, and the blue were left, presumably because the birds were unable to perceive them. Again, when the birds were given a patch of rice grains of which half was feebly illuminated by red light and the other half more strongly by blue light, they took the red but left the blue. Previous experiment had shewn that with ordinary white light the birds always started on the best illuminated grains. It seems reasonable to conclude, therefore, that in the red-blue experiment the feebly illuminated red grains were more visible than the far more strongly lighted blue ones. It might be objected that the birds had a prejudice against blue, but, as Hess points out, this is almost certainly not the case because they took grains

[1] C. Hess, *Handbuch der vergleichenden Physiologie* (herausgegeben von H. Winterstein), Bd. 4, 1912, p. 563.

which were very strongly illuminated with blue. Results of a similar nature were also obtained from pigeons, and from a kestrel which was fed with pieces of meat lighted with different colours.

On the whole these experiments of Hess convey a strong suggestion that the colour perceptions of birds may be quite different from our own, more especially where blue is concerned. Great caution is needed in discussing instances of mimicry in their relation to the bird, for we have no right to assume that the bird sees things as we do. On the other hand, it is a matter of much interest to find that in general blue plays relatively little part in cases of mimetic resemblance among butterflies; some combination of a dark tint with either red, white, brown, or yellow being far more common.

It will probably be admitted by most people that the evidence, taken all together, is hardly sufficient for ascribing to birds that part in the establishing of a mimetic likeness which is required on the theory of mimicry. That birds destroy butterflies in considerable numbers is certainly true, but it is no less true that some of the most destructive birds appear to exercise no choice in the species of butterfly attacked. They simply take what comes first and is easiest to catch. It is probably for this reason that the Wood-swallow feeds chiefly on Euploeines and Danaines (cf. p. 112). It is probably for this reason also that such a large proportion of the records of attacks on butterflies under natural conditions refer to the Pierines; for owing to their light colour it is probable that the "Whites" are more conspicuous and offer a better mark for a bird in pursuit than darker coloured species.

Mammals. Apart from man it is clear that only such mammals as are of arboreal habits are likely to cause destruction among butterflies in the imago state. Apparently there are no records of any arboreal mammal, except monkeys, capturing butterflies in the wild state, nor is there much evidence available from feeding experiments. But such evidence as exists is of considerable interest. As the result of feeding butterflies of different sorts to an Indian Tree-shrew (*Tupaia ferruginea*) Finn[1] found that it shewed a strong dislike to Danaids and to *Papilio aristolochiae* though it took readily *Papilio demoleus*, *Neptis kamarupa*, and *Catopsilia* (a Pierine). It is fairly certain that if the Tree-shrew is an enemy of butterflies in the wild state it is a discriminating one.

The other mammals with which experiments have been made are the common baboon, a monkey (*Cercopithecus pygerythrus*), and a mongoose (*Herpestes galera*)—all by Marshall[2] in South Africa. The mongoose experiments were few and inconclusive, nor is this a matter of much moment as it is unlikely that this mammal is a serious enemy of butterflies.

The monkey ate various forms of *Precis* (a Vanessid), after which it was given *Acraea halali*. This distasteful form was "accepted without suspicion, but when

[1] *Journ. As. Soc. Bengal*, vol. 66[2], 1898.
[2] *Trans. Ent. Soc. Lond.* 1902.

the monkey put it into his mouth, he at once took it
out again and looked at it with the utmost surprise
for some seconds, and then threw it away.   He would
have nothing to do with an *Acraea caldarena* which I
then offered him[1]."

The experiments with the baboons were more extensive.   Two species of *Acraea, halali* and *axina*,
were recognised when first offered and refused un-
tasted.   *Danais chrysippus*, on the other hand, was
tasted on being offered for the first time, and then
rejected.   This species was twice offered subsequently
and tasted each time before being rejected.   When
offered the fourth time it was rejected at sight.   The
baboon evidently learned to associate an unpleasant
taste with the *chrysippus* pattern.   At this stage it
would have been interesting to have offered it some
well-known mimic of *chrysippus*, such as the female
of *Hypolimnas misippus* or the *trophonius* form of
Papilio dardanus, but this experiment was unfortunately
not made.   Marshall did, however, offer it at the same
time a specimen each of *Byblia ilithyia* (a Vanessid)
and of *Acraea axina* to which it bears a general
resemblance.   The baboon took the former but ne-
glected the latter altogether.   The general resemblance
between the two species was not sufficiently close to
deceive it.

These experiments with mammals, though few in
number, are of unusual interest.   Should they be
substantiated by further work it is not impossible

[1] Marshall, *loc. cit.* p. 379.

that, as a factor in the establishing of a mimetic likeness, a stronger case may be made out for the monkey
than the bird.   The monkey apparently eats butterflies
readily[1]: owing probably to a keener sense of smell it
shews far less hesitation as to its likes and dislikes:
its intelligence is such that one can easily imagine it
exercising the necessary powers of discrimination;
in short it is the ideal enemy for which advocates of
the mimicry theory have been searching—if only it
could fly.   As things are its butterfly captures must be
made when the insect is at rest, probably near sunrise
and sunset, and this leads to a difficulty.   Most butterflies rest with their wings closed.   In many of the
well-known cases of mimicry the pattern on the under
surface of the mimic's wings which would meet the
monkey's eye is quite different from that of its model.
It is difficult in such cases to imagine the monkey
operating as a factor in establishing a resemblance
between the upper surfaces of the wings of the two
unrelated species.   On the other hand, some butterflies,

[1] In this connection may be quoted a letter from Capt. N. V. Neal
near Lagos to Mr W. A. Lamborn which was recently published in the
Proceedings of the Entomological Society.

"You have asked me about monkeys eating butterflies. This is
very common, as every native will tell you. I have seen it myself.
The monkey runs along a path, sees some butterflies fluttering round
some filth, goes very quietly, and seizes one by the wings, puts the
solid part (body) into his mouth, then pulls the wings off. The poor
butterfly goes down like any oyster....The dog-faced baboon and the
large brown monkey with a very long tail, which seems to be the most
common species in this colony, are great butterfly-eaters. The little
spider-monkey also considers a butterfly a treat, and prefers one to
a spider."

e.g. Papilio polytes, rest with wings outspread, and there are rare cases, such as that of *P. laglaizei* (p. 27), where the most striking point about the resemblance is only to be appreciated when the insects are at rest with their wings closed.   In such cases it is conceivable that the monkey may play a part in the elimination of the non-mimetic elements of a palatable species which at the same time possessed a mimetic form closely resembling another species disagreeable to the monkey's taste.   As has been pointed out earlier (p. 96) even a slight persecution directed with adequate discrimination will in time bring about a marked result where the mimetic likeness is already in existence.   It is not impossible therefore that the establishing of such a likeness may often be due more to the discrimination of the monkey than to the mobility of the bird.

CHAPTER X

MIMICRY AND VARIATION

I⊤ is clear from the last few chapters that the theory of mimicry in butterflies with its interpretation of the building up of these likenesses by means of natural selection in the form of predaceous birds and other foes is open to destructive criticism from several points of view. The evidence from mimicry rings makes it almost certain that in some cases the resemblance must be founded on an initial variation of such magnitude that the mimic could straightway be confused with the model. Till the mimic can be mistaken for the model natural selection plays no part. The evidence from breeding suggests strongly that in certain cases (*e.g. Papilio polytes*) the likeness arose in the form in which we know it to-day. In such cases there is no reason for supposing that natural selection has had anything to do with the formation of the finished mimic. Considerations of this nature may be said to have destroyed the view, current until quite recently, that in the formation of a mimetic resemblance the exclusive agent was natural selection. During the past few years it has come to be admitted by the staunchest upholders of the theory of mimicry that natural selection would not come into play until the would-be mimic was sufficiently like the model to be confused with it under natural conditions[1]. The part now often attributed to natural selection is to put a polish on the resemblance and to keep it up to the mark by weeding out those which do not reach the required standard. It is supposed that if natural selection ceases to operate the mimetic resemblance is gradually lost owing to the appearance of variations which are no longer weeded out. An interesting case has recently been brought forward by Carpenter[2] and explained on these lines: The Nymphaline *Pseudacraea eurytus* is a polymorphic species found in Central Africa. In Uganda it occurs in several distinct forms which were originally supposed to be distinct species. Three of these forms bear a marked resemblance to three species of the Acraeine genus *Planema*.

Mimic	Model
Pseudacraea eurytus	*Planema*
Form *hobleyi*[3] (Pl. VII, figs. 6, 7)	*macarista* (Pl. VII, fig. 2)
terra (Pl. VII, fig. 8)	*tellus* (Pl. VII, fig. 3)
obscura	*paragea* (Pl. VII, fig. 4)

These different species occur round Victoria Nyanza and also on some of the islands in the lake. Some

[1] Cf. E. B. Poulton in *Bedrock* for Oct. 1913, p. 301.

[2] *Trans. Ent. Soc. London*, 1914.

[3] In the female *hobleyi*, with rare exceptions, the orange of the male is replaced by white, and it has received the name *tirikensis*. The female of *P. macarista* also shews white in place of the orange of the male.

interesting points are brought out by a comparison of the occurrence and variation of the species on the mainland with what is found on Bugalla Island in the Sesse Archipelago. On the mainland the Pseud-acraeas are abundant but the Planemas even more so, outnumbering the former by about 5 : 2[1]. Moreover, it is rare to find individuals more or less intermediate between the three forms, though they are known to occur. On Bugalla Island, however, a different state of things is found. The Pseudacraeas are very abund-ant, whereas the Planemas, owing doubtless to the scarcity of their food plant, are relatively rare, and are very greatly outnumbered by the Pseudacraeas. At the same time the proportion of transitional forms among the Pseudacraeas is definitely higher than on the mainland. These facts are interpreted by Car-penter as follows:—

On the mainland where the models are abundant there is a vigorous action on the part of natural selection. The mimetic forms have a strong advantage and the non-mimetics have been gradually weeded out. But on the island, where the Pseudacraeas outnumber the models, the advantage obtained through mimicry is not so great. The so-called transitional forms are little, if at all, worse off than those closely resembling the scarce models, and consequently have as good a chance of surviving as any of the typical mimetic forms. On

[1] Cf. Poulton, E. B., I*er* Congr. Internat. d'Entomol., Bruxelles 1911. This proportion is founded on several hundreds caught at random. Observers are agreed that *Pseudacraea* is both a warier insect and a stronger flyer than the various Planemas which it resembles.

the mainland, however, the enemies of *Pseudacraea* are well acquainted with the *Planema* models which are here common, and discriminate against individuals which are not close mimics of the Planemas. The result is that on the mainland transitional forms are scarcer than on the island. Natural selection maintains a high standard for the mimetic likeness on the mainland owing to the abundance of the model; but when the model is scarce the likeness ceases to be kept up to the mark strictly, and tends to become lost owing to the appearance of fresh variations which are no longer weeded out.

Here it should be stated that the various Pseudacraeas form a population in which the different forms mate freely with one another. In the few breeding experiments that Dr Carpenter was able to make he found that *obscura* could produce *terra*, and that *tirikensis* was able to give *obscura*, the male in each case being, of course, unknown. Far too little work has as yet been done on the genetics of these various forms, and it would be rash to make assumptions as to the nature of the intermediates until the method of experimental breeding has been more extensively employed in analysing their constitution. Possibly it is not without significance that the abundance or scarcity of the *obscura* form runs parallel with the abundance or scarcity of the intermediates. It suggests that the intermediates are heterozygous in some factor for which the typical *obscura* is homozygous, and the fact that the intermediates are more numerous than *obscura* is what is to be looked for in a population mating at random.   This case of the polymorphic *Pseudacraea eurytus* is one of the greatest interest, but it would be hazardous to draw any far-reaching deductions from such facts as are known at present.   When the genetics of the various typical forms and of the intermediates has been worked out it will be disappointing if it does not throw clear and important light on these problems of mimetic resemblance.

As the result of modern experimental breeding work it is recognised that an intermediate form between two definite varieties may be so because it is hetero-zygous for a factor for which one variety is homozygous and which is lacking in the other—because it has received from only one parent what the two typical varieties receive from both parents or from neither. Its germ cells, however, are such as are produced by the two typical forms, and the intermediate cannot be regarded as a stage in the evolution of one variety from the other.   In these cases of mimicry the existence of intermediate forms does not entail the deduction that they have played a part in the evolution of one pattern from another under the influence of a given model. It is quite possible that the new mimetic pattern appeared suddenly as a sport and that the intermediates arose when the new form bred with that which was already in existence.   But before we are acquainted with the genetic relationships between the various forms, both types and intermediates, speculation as to their origin must remain comparatively worthless.

In this connection a few words on another source of variation may not be out of place. The patterns of butterflies are often very sensitive to changes in the conditions to which they are exposed during later larval and pupal life. Many moths and butterflies in temperate climates are double brooded. The eggs laid by the late summer brood hatch out, hibernate in the larval or pupal state, and emerge in the following spring. This spring brood produces the summer brood during the same year. In these cases it often happens that the two broods differ in appearance from one another, a phenomenon to which the term "Seasonal Dimorphism" has been applied. A well-marked instance is that of the little European Vanessid, *Araschnia levana*. The so-called *levana* form which emerges in the spring is a small black and orange-brown butterfly (Pl. VI, fig. 10). From the eggs laid by this brood is produced another brood which emerges later on in the summer, and is, from its very different appearance, distinguished as the *prorsa* form (Pl. VI, fig. 9). It is very much darker than the spring form and is characterised by white bands across the wings. The eggs laid by the *prorsa* form give rise to the *levana* form which emerges in the following spring. It has been shewn by various workers, and more especially by the extensive experiments of Merrifield[1], that the appearance of the *levana* or the *prorsa* form from any batch of eggs, whether laid by *prorsa* or *levana*, is dependent upon the conditions of temperature under

[1] *Ier Congr. Internat. d'Entom.*, Bruxelles 1911.

which the later larval and early pupal stages are passed.
By cooling appropriately at the right stage *levana* can
be made to produce *levana* instead of the *prorsa* which
it normally produces under summer conditions. So
also by appropriate warming *prorsa* will give rise to
prorsa. Moreover, if the conditions are properly ad-
justed an intermediate form *porima* can be produced,
a form which occurs occasionally under natural conditions. The pattern is, in short, a function of the
temperature to which certain earlier sensitive stages
in this species are submitted. What is true of *A. levana*
is true also of a number of other species. In some
cases temperature is the factor that induces the variation. In other countries where the year is marked
by wet and dry seasons instead of warm and cold ones
moisture is the agent that brings about the change.
In some of the South African butterflies of the genus
Precis the seasonal change may be even more conspicuous than in *A. levana*. In *Precis octavia*, for
example, the ground colour of the wet season form is
predominantly red, while in the dry season form of
the same species the pattern is different, blue being
the predominating colour (cf. Pl. VI, figs. 11 and 12).
Such examples as these are sufficient to shew how
sensitive many butterflies are to changes in the conditions of later larval and earlier pupal life. The
variations brought about in this way are as a rule
smaller than in the examples chosen, but in no case
are they known to be inherited, and in no case consequently could variation of this nature play any part in

9—2

evolutionary change. Before any given variation can be claimed as a possible stage in the development of a mimetic likeness satisfactory evidence must be forthcoming that it is not of this nature, but that it is transmissible and independent of climatic and other conditions.

Many species of butterflies, especially such as are found over a wide range, exhibit minor varieties which are characteristic of given localities. These minor varieties may be quite small. In *Danais chrysippus*, for example, African and Asiatic specimens can generally be distinguished. On examples from India a small spot is seen just below the bar on the fore wing and on the inner side of it. Eastwards towards China this spot tends to become larger and confluent with the white bar, giving rise to an L-shaped marking; westwards in Africa the spot tends to disappear altogether. The existence of such local races has been used as an argument for the hereditary transmission of very small variations—in the present instance the size of a small white spot[1]. For if it can be supposed that small differences of this nature are always transmitted, it becomes less difficult to imagine that a mimetic resemblance has been brought about by a long series of very small steps. But before this can be admitted it is necessary to shew by experiment that the size of this spot is independent of environmental conditions, both climatic and other. Apart from temperature and moisture it is not improbable that the formation of

---

[1] Cf. Poulton, *Bedrock*, Oct. 1913, p. 300.

pigment in the wings may depend in some degree upon the nature of the food. The larvae of *D. chrysippus* feed upon various Asclepiads, and it is at any rate conceivable that the pigment formation, and consequently the details of pattern, may be in slight measure affected by the plant species upon which they have fed. The species of food plants are more likely to be different at the extremities of the range of a widely distributed form like *D. chrysippus*, and if they are really a factor in the pattern it is at the extremities that we should expect to find the most distinct forms[1]. Actually we do find this in *D. chrysippus*, though it does not, of course, follow that the cause suggested is the true one, or, if true, the only one. Of the nature of local races too little at present is known to enable us to lay down any generalization. We must first learn by experiment how far they remain constant when transported from their own environment and bred in the environment under which another distinct local race is living. The behaviour of the transported race under the altered conditions would help us in deciding whether any variation by which it is characterised had a definite hereditary basis or was merely a fluctuation dependent upon something in the conditions under which it had grown up. The decision as to whether it is hereditary or not must depend upon the

---

[1] The size of the white spot may shew much variation in specimens from the same region. I have seen African specimens in which it is large, while in the Ceylon specimen figured on Plate IV it is as small as in the typical African specimen shewn on Plate VIII.

test of breeding, through which alone we can hope to arrive at a satisfactory verdict upon any given case.

The particular geographical variation which has just been considered happens to be a small one. But it may happen that a geographical variety is much more distinct. Indeed it is not impossible that butterflies which are at present ranked as distinct species may prove eventually to be different forms of the same species. Especially is this likely to be true of many forms in South America, of which Bates long ago remarked "that the suspicion of many of the species being nothing more than local modifications of other forms has proved to be well founded." Since Bates' day more material has been forthcoming[1] and it has been shewn that certain colour schemes are characteristic of distinct geographical regions in South America where they may occur in species belonging to very different genera and families. In Central America, for example, the pattern common to many species is determined by horizontal and oblique black bands on a bright fulvous brown ground, with two broken yellow bars towards the tip of the fore wing. The general type is well shewn by *Mechanitis saturata* and the female of *Dismorphia praxinoe* (Pl. X, figs. 7 and 3). Belonging to this pattern group are a number of different species belonging to various families, including several Heliconines and Ithomiines, Pierids such as *Dismorphia* and *Perrhybris*, Nymphalines of the genera *Eresia* and

---

[1] See Moulton, J. C., *Trans. Ent. Soc. London*, 1909.

Protogonius, and other forms. In Eastern Brazil the predominant pattern is one characterised by a yellow band across the hind wing and a white or yellow apical fore wing marking (cf. Pl. XV, figs. 3 and 8). Here also, with the exception of the *Perrhybris*, all the various genera which figured in the last group are again represented. It is true that the members of this second group are regarded as belonging to different species from those of the first group, but as species here are made by the systematist chiefly, if not entirely, on the colour pattern this fact may not mean much. Passing now to Ega on the Upper Amazons the general ground colour is a deep chestnut purple and the apical area of the fore wings presents a much mottled appearance (cf. Pl. XV, figs. 4 and 9). In this group again we find represented the different genera found in the other groups, the only notable absentees being *Eresia* and *Perrhybris*. Lastly in Ecuador, Peru, and Bolivia the general pattern scheme consists of orange-tawny markings on a black ground (cf. Pl. XV, figs. 5 and 10). This group differs somewhat in composition from the preceding in that it contains no Pierid and no Danaid. On the other hand its numbers have been strengthened by the accession of a *Papilio*, an *Acraea*, and two species of the Satyrid genus *Pedaliodes*. Certain writers have seen in the theory of mimicry the only explanation of these peculiar geographical pattern groups. The fashion is in each case set by the most abundant form, generally an Ithomiine of the genus *Melinaea*. The rest are mimics of this dominant species, either in the

Batesian or Müllerian sense.   Batesian mimics are such
genera as *Dismorphia* and *Protogonius,* to which there
are no reasons for attributing disagreeable properties.
Of the nature of Müllerian mimics on the other hand are
the various Heliconines and Ithomiines which enter
into the combination.   In each case the whole assemblage is a great "mimicry ring," of which the pattern
is dictated by the Ithomiine that predominates in point
of numbers.   It is, however, very doubtful whether
this can be accepted as a satisfactory explanation.   The
four groups which we have considered are all characterised by a peculiar and distinctive coloration, and in
each case the pattern must on the theory of mimicry
be regarded as a highly efficient warning pattern.   One
or other of these patterns must doubtless be looked upon
as the most primitive.   If so the question at once
arises as to why a distasteful genus should change from
one efficient warning pattern to another quite distinct
one.   If the newer pattern affords better protection
we should expect it to have spread and eventually to
have ousted the older one.   That it has not done so
must probably be attributed to the old pattern being
as efficient as the new one.   But if this is so we are
left without grounds for assuming the change to have
been brought about by natural selection through the
agency of enemies to whom warning colours appeal.
For natural selection can only bring about a change
that is beneficial to the species.   Hence we must
suppose the change on the part of the dominant model
to have been independent of natural selection by

enemies, and due to some condition or set of conditions of which we are ignorant. It is not inconceivable that the new colour scheme was associated with some physiological peculiarity which was advantageous to the species in its altered surroundings. If so natural selection may have favoured the new variety, not because of its colour scheme, but owing to the under-lying physiological differences of which the pattern is but an outward sign. And if this could happen in one species there seems to be no reason why it should not happen in others. The weak point of the explanation on the mimicry hypothesis is that it offers no explana-tion of the change in the so-called dominant Ithomiine pattern as we pass from one region to another. What-ever the cause of this change may be there would appear to be nothing against it having also operated to produce similar changes in other unrelated species, in which case the mimicry hypothesis becomes super-fluous. It is not unlikely that the establishing of these new forms was due to natural selection. If they were associated with physiological peculiarities better adapted for their environment it is reasonable to suppose that natural selection would favour their persistence as opposed to the older type until the latter was elimi-nated. But such action on the part of natural selection is quite distinct from that postulated on the mimicry hypothesis. On the one view the colour itself is selected because it is of direct advantage to its possessor; on the other view the colour pattern is associated with a certain physiological constitution which places the butterflies possessing it at an advantage as compared with the rest[1].

It is, nevertheless, possible that mimicry may have played some part in connection with establishing the new colour pattern in some of these South American species. For if the new pattern had become established in the predominant distasteful species, and if some of the members of a palatable form (*e.g. Protogonius*) were to shew a variation similar to that already established in the distasteful species, and if further there be granted the existence of appropriate enemies, then it would be almost certain that the newer form in palatable species would eventually replace the older form. In such a case the part played by natural selection would be the preservation of a chance sport which happened to look like an unpalatable form. There is no reason for regarding the change as necessarily brought about by the gradual accumulation of a long series of very small variations through the operation of natural selection.

[1] In this connection it is of interest that a recent observer with considerable breeding experience finds that the dark *doubledayaria* variety of the Peppered Moth is more hardy than the typical form (cf. p. 101). The swift success of the dark variety led some to regard it as better protected against bird enemies. It is, however, not unlikely that the deeper pigmentation is associated with some physiological difference which makes for greater hardiness. See Bowater, *Journal of Genetics*, vol. 3, 1914.

CHAPTER XI

FROM the facts recorded in the preceding chapters it is clear that there are difficulties in the way of accepting the mimicry theory as an explanation of the remarkable resemblances which are often found between butterflies belonging to distinct groups. Of these difficulties two stand out beyond the rest, viz., the difficulty of finding the agent that shall exercise the appropriate powers of discrimination, and the difficulty of fitting in the theoretical process involving the incessant accumulation of minute variations with what is at present known of the facts of heredity.

With regard to the former of these two difficulties we have seen that the supporters of the theory regard birds as the main selective agent. At the outset we are met with the fact that relatively few birds have been observed to prey habitually on butterflies, while some at any rate of those that do so shew no discrimination between what should be theoretically pleasant to eat and what should not be pleasant. Even if birds are the postulated enemies it must be further shewn that they exercise the postulated discrimination. It is required of them that they should do two things.

In the first place they must confuse an incipient or "rough" mimic with a model sufficiently often to give it an advantage over those which have not varied in the direction of the model. In other words, they must be easily taken in. Secondly, they are expected to bring about those marvellously close resemblances that sometimes occur by confusing the exact mimicking pattern with the model, while at the same time eliminating those which vary ever so little from it. In other words, they must be endowed with most remarkably acute powers of discrimination. Clearly one cannot ask the same enemy to play both parts. If, therefore, birds help to bring about the resemblance we must suppose that it is done by different species—that there are some which do the rough work, others which do the smoothing, and others again which put on the final polish and keep it up to the mark. This is, of course, a possibility, but before it can be accepted as a probability some evidence must be forthcoming in its favour.

But even if the difficulty of the appropriate enemy be passed over, and it be granted that a mimetic resemblance can be built up through a number of small separate steps, which have become separately established through the agency of separate species of birds with various but distinct discriminating powers, we are left face to face with an even more serious physiological difficulty. For why is it that when the end form which is supposed to have resulted from this process is crossed back with the original form all the intermediate steps do not reappear? Why is it that when the altered germplasm is mingled with the original germplasm the various postulated stages between them are not reformed? For in various cases where we know the course of evolution this does occur. The pale pink sweet-pea has come from the wild purple by a series of definite steps, and when it is crossed back with the wild form the resulting plants give the series of stages that have occurred in the evolution of the pink. So also when the orange rabbit is crossed with the wild grey form and the offspring are inbred there are reproduced the black, the tortoiseshell, and the chocolate, forms which are stages in the evolution of the orange from the wild grey. If then, to take an example, the "aristolochiae" form of *Papilio polytes* has been derived from the male-like form by a series of steps, why do we not get these steps reproduced after the germplasms of the two forms have been mingled? From the standpoint of modern genetic work the inference is that these postulated intermediate steps have never existed—that the one form of *polytes* female came directly from the other, and was not built up gradually through a series of stages by the selective agency of birds or any other discriminating enemy.

These two objections, viz. the difficulty of finding the appropriate enemy, and the non-appearance of intermediates when the extreme forms are crossed, may, perhaps, be said to constitute the main objections to the current theory of mimicry. Others such as the relative scarcity of mimicry in the male sex and the existence of cases of polymorphism among females of a species which cannot possibly be explained on mimetic lines have already been mentioned. But while the main objections remain it is hardly necessary to insist upon these others. Looked at critically in the light of what we now know about heredity and variation the mimicry hypothesis is an unsatisfactory explanation of the way in which these remarkable resemblances between different species of butterflies have been brought about. Sometimes this is admitted by those who nevertheless embrace the theory with a mild aloofness. For they argue that even though it does not explain all the facts no other theory explains so many. Others have sought an explanation in what has sometimes been termed the hypothesis of external causes, regarding these resemblances as brought about by similar conditions of soil and climate, and so forth. It is not inconceivable that certain types of colour and pattern may be the expression of deep-seated physiological differences, which place their possessors at an advantage as compared with the rest of the species. Were this so it is but reasonable to suppose that they would become established through the agency of natural selection. But it is difficult, if not impossible, to regard this as a satisfactory solution, if for no other reason than that it offers no explanation of polymorphism. For example, each of the three forms of *polytes* female holds its own and all must, therefore, be regarded as equally well adapted to the circumstances under which they live. They are so distinct in colour that it is difficult on this hypothesis to suppose that they are all on the same footing in respect to their environment. Yet if one is better off than the others, how is it that these still exist?

Those who have examined long series of these cases of resemblance among butterflies find it hard to believe that there is not some connection between them apart from climatic influence. One feels that they are too numerous and too striking to be all explained away as mere coincidences engendered by like conditions. Nor is it improbable that natural selection in the form of the discriminating enemy may have played a part in connection with them, though a different one from that advocated on the current theory of mimicry. If we assume that sudden and readily appreciable variations of the nature of "sports" turn up from time to time, and if these variations happen to resemble a form protected by distastefulness so closely that the two can be confused by an enemy which has learned to avoid the latter, then there would appear to be good grounds for the mimicking sport becoming established as the type form of the species. For it has already been seen that a rare sport is not swamped by intercrossing with the normal form, but that on the contrary if it possess even a slight advantage, it must rapidly displace the form from which it sprang (cf. Chap. VIII). On this view natural selection in the form of the discriminating enemy will have played its part, but now with a difference. Instead of building up a mimetic likeness bit by bit it will merely have conserved and rendered numerically preponderant a likeness which had turned up quite independently. The function of natural selection in respect of a mimetic likeness lies not in its formation but in its conservation. It does not bring about the likeness, neither does it accentuate it: it brings about the survival of those forms which happen to shew the likeness. Why variations on the part of one species should bear a strong resemblance to other, and often distantly related, species is another question altogether.

Even a superficial survey of the facts makes it evident that cases of mimicry tend to run in series—that a closely related series of mimics, though often of very different pattern and colour, tends to resemble a closely related series of models. In Asia we have the Cosmodesmus Papilios mimicking a series of Danaines, while the true Papilios (cf. Appendix II) tend to resemble a series of the less conspicuous members of the Pharmacophagus group. In the same region the various species of *Elymnias* form a series resembling a series of Danaines. In Africa there stands out the Cosmodesmus group again mimicking a Danaine series, and in part also an Acraeine series. Overlapping the Acraeines again are various forms of the Nymphaline genus *Pseudacraea*. It is also of interest that in *Danais chrysippus* and *Acraea encedon* the Danaine and Acraeine series overlap (cf. Pl. IX). Similar phenomena occur also in South America, where closely parallel series of colour patterns are exhibited by several

Ithomiines, by *Heliconius, Lycorea, Dismorphia,* and other genera (cf. p. 39). On the other hand such mimetic resemblances as are shewn by the South American Swallow-tails of the Papilio and Cosmodesmus groups are almost all with the Pharmacophagus group, and almost all of the red-black kind (cf. p. 43).

On the whole it may be stated that the majority of cases of mimicry fall into one or other of such series as the above. If we select a case of mimicry at random we shall generally find that there are at least several close allies of the mimic resembling several close allies of the model. Isolated cases such as the resemblance between *Pareronia* and *Danais* (p. 23), between *Archonias* and a Pharmacophagus Papilio (p. 43), or the extraordinary instance of *Papilio laglaizei* and *Alcidis agathyrsus,* must be regarded as exceptional.

We have before us then a number of groups of butterflies each with a series of different colour patterns. In each group a portion of the series overlaps a portion of the series belonging to another more or less distantly related group. In the light of recent discoveries connected with heredity and variation the natural interpretation to such a set of phenomena would be somewhat as follows: Each group of Lepidoptera, such as those just discussed, contains, spread out among its various members, a number of hereditary factors for the determination of colour pattern. Within the group differences of pattern depend upon the presence or absence of this or that factor, the variety of pattern being the result of the many possible permutations and

combinations of these colour factors. Within the
limits of each group is found a definite number of these
factors—more in one group, less in another. But some
factors may be common to two or more groups, in
which case some of the permutations of the factors
would be similar in the groups and would result in
identical or nearly identical pattern. To take a simple
example in illustration, let us suppose that a given
group, (a), contains the eight factors A—H. Since
any species in the group may exhibit any combination
of one or more of these factors it follows that a considerable number of different forms are possible. Now
suppose that another group, (β), distinguished from
(a) by definite structural features, also contains eight
factors within the group, and that these factors are
F—M, F, G, and H being common to both (a) and
(β). Any combination therefore in (a) lacking the
factors A—E will be paralleled by any combination in
(β) lacking the factors I—M. For in both cases we
should be dealing only with the factors F, G, and H,
which are common to each group. So again a third
group might have some factors in common with (a)
and some with (β), and so on for other groups. In
this way certain of the series of colour patterns found
in (β) would overlap certain of those in (a), while others
of the groups (β) and (a) might overlap those found in
different groups again. The striking resemblances not
infrequently found between species belonging to quite
distinct groups would on this view depend upon the
hereditary factors for pattern and colour being limited

in number, so that the same assortment might not infrequently be brought together even though the group whose members exhibited the resemblance might, owing to structural differences, be placed in different families.

We know from recent experimental work that something of the sort is to be found in the coat colours of different rodents. Agouti, black, chocolate, blue-agouti, blue, and fawn form a series of colours common to the rabbit, the mouse, and the guinea-pig. These colours are related to each other in the same way in these different beasts. In the rat, on the other hand, there occur of this range of colours only the agouti and the black. Each of these species again has certain colour patterns which are peculiar to itself, such as the "English" type in the rabbit, the tricolor pattern in the guinea-pig, or the "hooded" variety in the rat. The total range of colour and pattern is somewhat different for each species, but a few are common to them all. Moreover, there are others which are common to the mouse and the rabbit but are not found in the guinea-pig, and others again which may occur in the rabbit and the guinea-pig but have not been met with in the other two. In certain features the rabbit might be said to "mimic" the mouse, and in other features the guinea-pig. It is not, of course, suggested that the case of the butterflies is so simple as that of the rodents, but so far as we can see at present there would seem to be no reason why the explanation should not be sought along the same lines.

10—2

On this view the various colour patterns found among butterflies depend primarily upon definite hereditary factors of which the number is by no means enormous. Many of these factors are common to several or many different groups, and a similar aggregate of colour factors, whether in an Ithomiine, a Pierid, or a Papilio, results in a similar colour scheme. The likeness may be close without being exact because the total effect is dependent in some degree on the size and relative frequency of the scales and other structural features. In so far as pattern goes *Hypolimnas dubius* and *Amauris echeria* (Pl. VIII, figs. 7 and 8) are exceedingly close. But inspection at once reveals a difference in the quality of the scaling, giving to the *Hypolimnas*, where the black and yellow meet, a softness or even raggedness of outline, which is distinct from the sharper and more clear-cut borders of the *Amauris*. It is not unreasonable to suppose that these species carry identical factors for colour pattern, and that the differences by which the eye distinguishes them are dependent upon the minuter structural differences such as occur in the scaling. So the eye would distinguish between a pattern printed in identical colours on a piece of cretonne and a piece of glazed calico. Though pattern and colour were the same the difference in material would yield a somewhat different effect.

On the view suggested the occurrence of mimetic resemblances is the expression of the fact that colour pattern is dependent upon definite hereditary factors of which the total number is by no means very great.

As many of the factors are common to various groups of butterflies it is to be expected that certain of the colour patterns exhibited by one group should be paralleled by certain of those found in another group. That cases of resemblance should tend to run in parallel series in different groups is also to be expected, for in some groups the number of factors in common is likely to be greater than in other groups. In consonance with this view is the fact that where polymorphism occurs among the females of a mimicking species the models, though often widely different in appearance, are, as a rule, closely related. Some of the Asiatic Papilios, for instance, resemble Danaines, while others resemble Pharmacophagus Papilios. But although the polymorphism exhibited by the females of a given species may be very marked, we do not find one of them resembling a Danaine and another a Pharmacophagus Swallow-tail. The models of a polymorphic mimic are almost always closely related species[1].

In discussing the problems of mimicry more attention is naturally paid to groups which exhibit the phenomenon than to those which either do not do so, or else only do so to a very limited extent. Yet the latter may be of considerable interest. Among the Pieridae of the Old World the phenomenon of mimicry is very rare. *Pareronia* and *Aporia agathon* conform

---

[1] As examples may be mentioned *P. polytes, Hypolimnas misippus, H. dubius,* and *Pseudacraea hobleyi.* With the exception of the *planemoides* form it is true also for *P. dardanus,* the most polymorphic of them all.

closely to the common Danaid type represented by *Danais vulgaris* and other species, but apart from these none of the many Pierids in Asia resemble any of the recognised models. Africa is apparently destitute of Pierids which mimic species belonging to other groups. Yet no group of butterflies is more persecuted by birds. Of all the instances of bird attacks collected together by Marshall[1] more than one-third are instances of attacks upon this group alone. If birds are the agents by which mimetic likenesses are built up through the cumulative selection of small variations, how can the rarity or absence of mimetic Pierids in the Old World be accounted for? For the species of Pierids, like the species of other families, shew considerable variation, and if this process of selection were really at work one would expect to find many more Pierid mimics in these regions than actually occur. It is true that the white, yellow, and red pigments found in Pierids differ from those of other butterflies in being composed either of uric acid or of some substance closely allied to that body[2]. These substances are generally found between the two layers of chitin, of which the scale is composed, whereas the black pigment is intimately associated with the chitin of the scale itself. What is perhaps the principal factor in the formation of a mimetic likeness is the distribution of the black pigment with reference to the lighter pigments; and although the latter are chemically distinct

---

[1] *Trans. Ent. Soc. Lond.* 1909.
[2] Cf. F. G. Hopkins, *Phil. Trans. Roy. Soc.* 1895.

in the Pierids as compared with other butterflies, there would seem to be no reason why the same factors governing the distribution of black should not be common to members of different groups. A distribution of black pigment similar to that found in a model and its mimic may occur also in a non-mimetic ally of the mimic. *Dismorphia astynome*, for example, resembles the Ithomiine *Mechanitis lysimnia* (Pl. XV, fig. 8) both in the distribution of black as well as of yellow and bright brown pigments. A similar distribution of the black pigment is also found in *Dismorphia avonia*, but the yellow and bright brown of the other two species is here replaced with white. By a slight though definite alteration in chemical composition this white pigment could be changed into bright brown and yellow with the result that *D. avonia* would closely resemble *D. astynome* in its colour scheme and would in this way also become a mimic of *Mechanitis lysimnia*. Another good instance is that of the females of *Perrhybris demophile* and *P. lorena*, the former being black and white, whereas in the latter the white is replaced by yellow and bright brown, giving the insect a typical Ithomiine appearance[1]. Here again a definite small change in the composition of the pigment laid down in the scales would result in the establishing of a mimetic likeness where there would otherwise be not even a suggestion of it. It is in accordance with what we know to-day of variation

[1] Coloured representations of these two species will be found on Pl. 20 of Seitz, *Macrolepidoptera of the World, Fauna Americana.*

that such a change should appear suddenly, complete
from the start.   And if so there is no difficulty in supposing that it might be of some advantage to its
possessor through the resemblance to an unpalatable
form.   Even were the advantage but a slight one it
is clear from previous discussion (p. 96) that the new
variety would more or less rapidly replace the form
from which it had sprung.   With the continued
operation of natural selection the new form would
entirely supplant the original one, but it is not impossible that in some cases the selecting agent may be
removed before this result has been achieved.   In this
event the proportions of the new and the old form
would fall into a condition of equilibrium as in *P.
polytes* in Ceylon, until some other selective agent
arose to disturb the balance.   On this view natural
selection is a real factor in connection with mimicry,
but its function is to conserve and render preponderant
an already existing likeness, not to build up that likeness through the accumulation of small variations, as
is so generally assumed.   Recent researches in heredity
and variation all point to this restriction of the scope
of natural selection.   Hitherto an argument in favour
of the older view has been that derived from the study
of adaptation—of an apparent purpose, which, at first
sight, appears to be behind the manner in which
animals fit into their surroundings.   For many the
explanation of this apparent purpose has been found
in the process of natural selection operating gradually
upon small variations, accumulating some and rejecting

others, working as it were upon a plastic organism, moulding it little by little to a more and more perfect adaptation to its surroundings. On this view adaptation is easy to understand. The simplicity of the explanation is in itself attractive. But when the facts come to be examined critically it is evident that there are grave, if not insuperable, difficulties in the way of its acceptance. To outline some of these has been the object of the present essay. Though suggestions have been made as to the lines along which an explanation may eventually be sought it is not pretended that the evidence is yet strong enough to justify more than suggestions. Few cases of mimicry have as yet been studied in any detail, and until this has been done many of the points at issue must remain undecided. Nevertheless, the facts, so far as we at present know them, tell definitely against the views generally held as to the part played by natural selection in the process of evolution.

APPENDIX I

FOR the table on p. 155 I am indebted to the kindness of Mr H. T. J. Norton of Trinity College, Cambridge. It affords an easy means of estimating the change brought about through selection with regard to a given hereditary factor in a population of mixed nature mating at random. It must be supposed that the character depending upon the given factor shews complete dominance, so that there is no visible distinction between the homozygous and the heterozygous forms. The three sets of figures in the left-hand column indicate different positions of equilibrium in a population consisting of homozygous dominants, heterozygous dominants, and recessives. The remaining columns indicate the number of generations in which a population will pass from one position of equilibrium to another, under a given intensity of selection. The intensity of selection is indicated by the fractions $\frac{100}{50}$, $\frac{100}{75}$, etc. Thus $\frac{100}{75}$ means that where the chances of the favoured new variety of surviving to produce offspring are 100, those of the older variety against which selection is operating are as 75; there is a 25 % selection rate in favour of the new form.

The working of the table may perhaps be best explained by a couple of simple examples.

In a population in equilibrium consisting of homozygous dominants, heterozygous dominants and recessives the last named class comprises 2·8 % of the total: assuming that a 10 % selection rate now operates in its favour as opposed to the two classes of dominants—in how many generations will the recessive come to constitute one-quarter of the population? The answer is to be looked for in column B (since the favoured variety is recessive) under the fraction $\frac{100}{90}$. The recessive

Number of generations taken to pass from one position to another as indicated in the percentages of different individuals in left-hand column

Percentage of total population formed by old variety	Percentage of total population formed by the hybrids	Percentage of total population formed by the new variety	A. Where the new variety is dominant				B. Where the new variety is recessive			
			$\frac{100}{50}$	$\frac{100}{75}$	$\frac{100}{90}$	$\frac{100}{99}$	$\frac{100}{50}$	$\frac{100}{75}$	$\frac{100}{90}$	$\frac{100}{99}$
99·9	·09	·000	4	10	28	300	1920	5740	17,200	189,092
98·0	1·96	·008	2	5	15	165	85	250	744	8,160
90·7	9·0	·03	2	4	14	153	18	51	149	1,615
69·0	27·7	2·8	2	4	12	121	5	13	36	389
44·4	44·4	11·1	4	8	12	119	2	6	16	169
25·	50·	25·	10	17	18	171	2	4	11	118
11·1	44·4	44·4	36	68	40	393	2	4	11	120
2·8	27·7	69·0	170	333	166	1,632	2	6	14	152
·03	9·0	90·7	3840	7653	827	8,243	4	10	16	165
·008	1·96	98·0			19,111	191,002			28	299
·000	·09	99·9								

passes from 2·8 % to 11·1 % of the population in 36 generations, and from 11·1 % to 25 % in a further 16 generations—*i.e.* under a 10 % selection rate in its favour the proportion of the recessive rises from 2·8 % to 25 % in 52 generations.

If the favoured variety is dominant it must be borne in mind that it can be either homozygous or heterozygous—that for these purposes it is represented in the left-hand column by the hybrids as well as by the homozygous dominant. In a population in equilibrium which contains about 2 % of a dominant form, the great bulk of these dominants will be heterozygous, and the relative proportion of recessives, heterozygous, and homozygous dominants is given in the second line of the left-hand column.

Let us suppose now that we want to know what will be the percentage of dominants after 1000 generations if they form 2 % of the population to start with, and if, during this period, they have been favoured with a 1 % selection advantage. After 165 generations the proportion of recessives is 90·7, so that the proportion of dominants has risen to over 9 %; after 153 further generations the percentage of dominants becomes 27·7 + 2·8 = 30·5; after 739 generations it is 88·8 %, and after 1122 generations it is 69·0 + 27·7 = 96·7. Hence the answer to our question will be between 89 % and 97 %, but nearer to the latter figure than the former.

Mr Norton has informed me that the figures in the table are accurate to within about 5 %.

APPENDIX II

THE genus *Papilio* is a large and heterogeneous collection. It was pointed out by Haase[1] that it falls into three distinct sections, of which one—the Pharmacophagus section—provides those members which serve as models in mimicry; while in the other two sections are found mimics, either of Pharmacophagus Swallow-tails, or of models belonging to other groups. Though Haase's terms have not yet come into general use with systematists, there is little doubt that the genus *Papilio* as it now stands must eventually be broken up on these lines. To say that one species of *Papilio* mimics another is therefore somewhat misleading; for the differences between the Pharmacophagus group and the other two are such as to constitute at any rate generic distinction in other groups. For convenience of reference a table has been added in which the various Papilios mentioned in the text have been assigned to their appropriate sections, and referred to their respective models.

[1] *Untersuchungen über die Mimikry*, 1893.

Pharmacophagus	Papilio	Cosmodesmus
(POISON-EATERS)	(FLUTED SWALLOW-TAILS)	(KITE SWALLOW-TAILS)
Antennae without scales.	Antennae without scales.	Antennae scaled on upper side.
Outer ventral row of spines of tarsi not separated from the dorsal spines by a spineless longitudinal depression.	Outer ventral row of spines of tarsi separated from the dorsal spines by a spineless longitudinal depression.	As in Papilio.
Larva covered with short hairs — with fleshy tubercles but no spines.	Larva either smooth or with hard spiny tubercles. Third and fourth thoracic segments enlarged.	Larva with third thoracic segment enlarged (known only in a few species).
Pupa with row of well-marked humps on each side of abdomen.	Pupa wrinkled — generally with short dorsal horn. Humps if present very short.	Pupa short with long four-sided thoracic horn.
Larva feeds on *Aristolochia*.	Larva does not feed on *Aristolochia*.	As in Papilio.
	Abdominal margin of hind wing curved downwards forming a kind of groove. No scent organ.	Abdominal margin of hind wing bent over in ♂, and with scent organ in fold so formed.

LIST OF PAPILIONINE MIMICS

Pharmacophagus (MODEL)	Papilio (MIMIC)	Cosmodesmus (MIMIC)	(MODELS)
			ORIENTAL
hector (V. 6)	agestor (II. 3)		Caduga tytia (II. 2)
aristolochiae (V. 5)	clytia (I. 8)		Danais septentrionis (I. 3)
coon	,, var. dissimilis (I. 7)		Euploea core (I. 10)
polyxenus	mendax (II. 9)		,, rhadamanthus (II. 8)
	paradoxus		,, mulciber (II. 5)
	polytes ♀ (V. 4)		
	,, ♀ (V. 3)		
	memnon ♀		
	bootes (III. 6)	delesserti	Alcidis agathyrsus (III. 1)
	laglaizei (III. 2)	macareus	Ideopsis daos (III. 4)
		xenocles (I. 4)	Danais septentrionis (I. 3)
			,,
			AFRICAN
	dardanus ♀ (VIII. 2)		Danais chrysippus (VIII. 5)
	,, ♀ (VIII. 3)		Amauris niavius (VIII. 6)
	,, ♀ (VIII. 4)		,, echeria (VIII. 7)
	echerioides ♀		,, psyttalea
	cynorta ♀ (VII. 10)		Planema epaea (VII. 5)
	rex		Melinda formosa
		ridleyanus (VI. 6)	Acraea egina (VI. 7)
		leonidas (VI. 2)	Danais petiverana (VI. 1)
		brasidas (VI. 4)	Amauris hyalites (VI. 3)
			AMERICAN
hahneli (mimic)	zagreus (X. 8)		Methona confusa (XII. 1)
	bachus		various Heliconinae and Ithomiinae
	euterpinus		Heliconius melpomene (XI. 5)
various species (XII. 1, 2, 3)	hippason, etc.	pausanias (XI. 2)	,, sulphurea (XI. 1)
philenor (XVI. 1)	troilus (XVI. 2)	lysithous etc. (XIII. 4, 5, 6)	
	turnus ♀		
	asterius ♀		

PLATE I

ORIENTAL BUTTERFLIES

1. *Pareronia ceylonica*   ♀ } (Pieridae)
2.   ,,   ,,   ♂
3. *Danais septentrionis*   (Danainae)
4. *Papilio xenocles*   (Papilionidae)
5. *Hypolimnas bolina*   ♂ } (Nymphalinae)
6.   ,,   ,,   ♀
7. *Papilio clytia* var. *dissimilis* } (Papilionidae)
8.   ,,   ,,   var. *lankeswara*
9. *Elymnias singhala*   (Satyrinae)
10. *Euploea core*   (Danainae)

× ⅔

Plate I

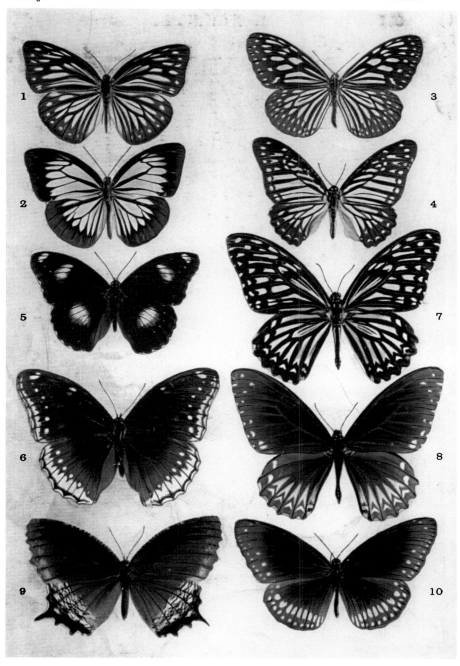

ORIENTAL BUTTERFLIES

PLATE II

ORIENTAL BUTTERFLIES

1. *Delias eucharis* (Pieridae)
2. *Caduga tytia* (Danainae)
3. *Papilio agestor* (Papilionidae)
4. *Euploea mulciber* ♂ ⎫
5.   ,,   ,, ♀ ⎬ (Danainae)
6. *Elymnias malelas* ♂ ⎫
7.   ,,   ,, ♀ ⎬ (Satyrinae)
8. *Euploea rhadamanthus* (Danainae)
9. *Papilio mendax* (Papilionidae)

ORIENTAL BUTTERFLIES

PLATE III

ORIENTAL MOTHS AND BUTTERFLIES

The three upper figures are those of moths, and the three lower ones are those of butterflies.

    1.   *Alcidis agathyrsus* (New Guinea)

    2.   *Papilio laglaizei*   ,,   ,,

The moth is here supposed to serve as a model for the far rarer Papilio.

    3.   *Cyclosia hestinioides*

    4.   *Ideopsis daos*

The butterfly is very common and must be regarded as the model, the rarer moth as the mimic.

    5.   *Epicopeia polydora* (Assam)

    6.   *Papilio bootes*   ,,

Both of these species are to be regarded as mimics of the abundant Pharmacophagus Papilio, *P. polyxenus*, which is very like *P. bootes* in appearance.

Plate III

$\times \frac{2}{3}$

ORIENTAL MOTHS AND BUTTERFLIES

PLATE IV

ORIENTAL BUTTERFLIES

1. *Danais chrysippus* ♂ ⎫
2. ,, *plexippus* ♀ ⎬ (Danainae)
3. *Argynnis hyperbius* ♀ ⎫
4. ,, ,, ♂ ⎬ (Nymphalinae)
5. *Elymnias undularis* ♀ ⎫
6. ,, ,, ♂ ⎬ (Satyrinae)
7. *Hypolimnas misippus* ♀ ⎫
8. ,, ,, ♂ ⎬ (Nymphalinae)

The two Danaids together with the females of the other three species form a "mimicry ring." For explanation see text, pp. 65–69.

ORIENTAL BUTTERFLIES

PLATE V

ORIENTAL BUTTERFLIES

1. *Papilio polytes* ♂
2. „   „   ♀, var. *cyrus* (*M* form)
3. „   „   ♀, var. *polytes* (*A* form)
4. „   „   ♀, var. *romulus* (*H* form)
5. „   *aristolochiae*
6. „   *hector*

The specimens figured on this plate were taken in Ceylon where they are all plentiful.

Figures 1*a*–6*a* represent the under surfaces of the hind wings belonging to specimens 1–6.

Plate V

ORIENTAL BUTTERFLIES

PLATE VI

AFRICAN BUTTERFLIES

(except *A. levana*, Figs. 8–10, which is European)

1. *Danais petiverana* (Danainae)
2. *Papilio leonidas* (Papilionidae)
3. *Amauris hyalites* (Danainae)
4. *Papilio leonidas* var. *brasidas* (Papilionidae)
5. *Pseudacraea boisduvali* (Nymphalinae)
6. *Papilio ridleyanus* (Papilionidae)
7. *Acraea egina* (Acraeinae)
8. *Araschnia levana* var. *porima*
9.   ,,   ,,   var. *prorsa*
10.   ,,   ,,
11. *Precis octavia* var. *sesamus*
12.   ,,   ,,   var. *natalensis*

AFRICAN BUTTERFLIES

PLATE VII

TROPICAL AFRICAN BUTTERFLIES

1. *Planema macarista* ♂ (Acraeinae)
2. „ „ ♀ „
3. „ *tellus* „
4. „ *paragea* „
5. „ *epaea* „
6. *Pseudacraea hobleyi* ♂ (Nymphalinae)
7. „ „ ♀ „
8. „ *terra* „
9. *Elymnias phegea* ♀ (Satyrinae)
10. *Papilio cynorta* ♀ (Papilionidae)

(NOTE. *Pseudacraea hobleyi* and *P. terra* (Figs. 6–8) were at one time regarded as separate species. More recently they have been shewn to be forms of the polymorphic species, *Pseudacraea eurytus*.)

TROPICAL AFRICAN BUTTERFLIES

PLATE VIII

AFRICAN BUTTERFLIES

1. *Papilio dardanus* ♂
2.   ,,   ,,   ♀, var. *trophonius*
3.   ,,   ,,   ♀, var. *hippocoon*
4.   ,,   ,,   ♀, var. *cenea*
5. *Danais chrysippus* (Danainae)
6. *Amauris niavius*   ,,
7.   ,,   *echeria*   ,,
8. *Hypolimnas dubius* var. *mima*   (Nymphalinae)
9.   ,,   ,,   var. *wahlbergi*   ,,

AFRICAN BUTTERFLIES

Plate IX

Danais chrysippus
a. Typical form
b. *Alcippus* form
c. *Dorippus* form

Acraea encedon
d. Typical form
e. *Alcippina* form
f. *Daira* form

Hypolimnas misippus ♀
g. Typical form
h. *Alcippoides* form
i. *Inaria* form

(After Aurivillius)

PLATE X

(NOTE. The figure of the *Mechanitis* (Fig. 7) is taken from a rather worn specimen. The quality of the orange brown is better shewn by the specimen illustrated in Fig. 7 on Plate XV.)

SOUTH AMERICAN BUTTERFLIES

PLATE XI

SOUTH AMERICAN BUTTERFLIES

1. *Heliconius sulphurea*   (Heliconinae)
2. *Papilio pausanias*   (Papilionidae)
3. *Heliconius telesiphe*   (Heliconinae)
4. *Colaenis telesiphe*   (Nymphalinae)
5. *Heliconius melpomene* (Heliconinae)
6. *Pereute charops*   ♀   (Pieridae)
7.   „   „   ♂   „
8. *Papilio osyris*   ♂   (Papilionidae)
9.   „   „   ♀   „
10. *Archonias critias* ♀   (Pieridae)

Plate XI

SOUTH AMERICAN BUTTERFLIES

Plate XII

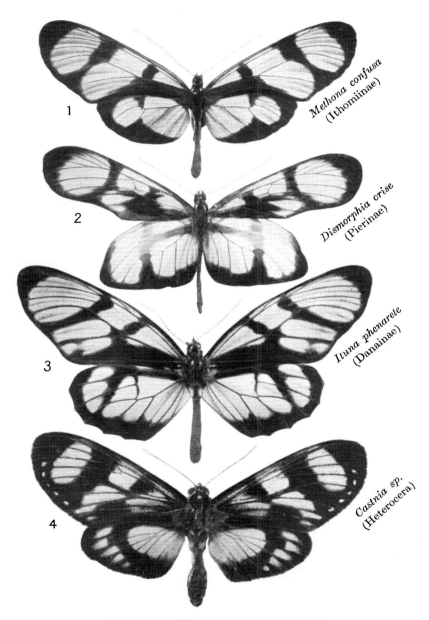

1

Methona confusa
(Ithomiinae)

2

Dismorphia crise
(Pierinae)

3

Ituna phenarete
(Danainae)

4

Castnia sp.
(Heterocera)

SOUTH AMERICAN BUTTERFLIES

Plate XIII

MODELS

1. *Papilio nephalion*
2.   ,,   *chamissonia*
3.   ,,   *perrhebus*

MIMICS

4. *Papilio lysithous* var. *lysithous*
5.   ,,   ,,   var. *rurik*
6.   ,,   ,,   var. *pomponius*

(For further details of this case see Jordan, *Ier Congr. Internat. d'Entomologie*, Bruxelles, 1911, p. 396.)

Plate XIV

1. *Methona confusa,* ×90
    (Ithomiinae)

2. *Dismorphia orise,* ×150
    (Pierinae)

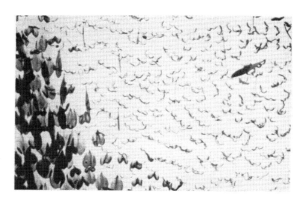

3. *Thyridia themisto,* ×90
    (Ithomiinae)

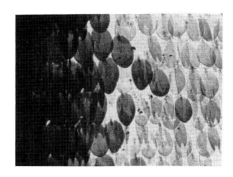

4. *Ituna ilione,* ×90
    (Danainae)

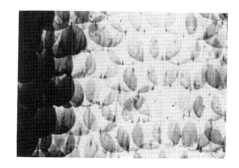

5. *Castnia sp.,* ×60
    (Moth)

Microphotographs of the scales of various Lepidoptera in the S. American
"Transparency group." For explanation see text, pp. 39–42.

PLATE XV

CENTRAL AND SOUTH AMERICAN BUTTERFLIES

Illustrating the closely parallel series of patterns occurring in the two distinct groups Heliconinae and Ithomiinae.

1.	*Heliconius*	*mirus*
2.	,,	*telchinia*
3.	,,	*eucrate*
4.	,,	*pardalinus*
5.	,,	*splendens*
6.	*Mechanitis*	*elisa*
7.	,,	*saturata*
8.	,,	*lysimnia*
9.	,,	*egaensis*
10.	,,	*methona*

CENTRAL AND SOUTH AMERICAN BUTTERFLIES

PLATE XVI

NORTH AMERICAN BUTTERFLIES

1. *Papilio philenor*   (Papilionidae)
2.   ,,   *troilus*   ,,
3. *Argynnis diana* ♀   (Nymphalinae)
4. *Limenitis arthemis*   ,,
5.   ,,   *astyanax*   ,,
6.   ,,   *archippus*   ,,
7.   ,,   *floridensis* (=*eros*)   ,,
8. *Danais archippus*   (Danainae)
9.   ,,   *berenice*   ,,

Plate XVI

NORTH AMERICAN BUTTERFLIES

INDEX

References to the plates are given in thicker type

Poison - eaters, see Pharmacophagus Swallow-tails
Polymorphism, in females of
mimicking species, 13; among
females of *P. dardanus*, 30;
among females of *P. polytes*, 75
Population, conditions of equilibrium in mixed, 93
Poulton, 17; on N. American
mimetic butterflies, 45; on the
"Transparency group," 41; on
mimicry through agency of
migratory birds, 53; on *Hypolimnas misippus*, 66 note; on
the relation between mimetic
forms of *P. polytes*, 90; on
predaceous insects, 105; on
relative proportion of different
forms of *Pseudacraea eurytus*,
127; on local variation in
D. chrysippus, 132
Precis, 111, 122, 131; *P. octavia*,
seasonal dimorphism in, 131,
VI. 11, 12
Prioneris, 110; *P. sita*, 28
Pritchett, feeding experiments
with lizards, 108
Protective resemblance, 8
Protogonius, as mimics of Ithomiines, 38; as members of
mimicry rings, 134, 135, 138;
P. tithoreides, **X. 9**
Pseudacraea, 59, 144; *P. boisduvali*, 34, **VI. 5**; *P. eurytus*,
relative proportion of different
forms in, 127; polymorphism
of in relation to model, 149
note; var. *hobleyi* as mimic of
Planema macarista, 35, 127,
VII. 6, 7; var. *terra*, as mimic
of *Planema tellus*, 126, **VII. 8**;
var. *obscura* as mimic of *Planema paragea*, 126

Ray, on adaptation, 4, 6
Rodents, bearing on mimicry of
recent genetic work with, 147

Satyrinae, transparency in S.
American, 42
Sceleporus floridanus, 108
Schaus, on birds as enemies of
butterflies, 112
Seasonal dimorphism, 130
Seitz, 44, 52, 58
Shelford, 56 note
S. American butterflies, mimicry
among, 38
Sports, as foundation of mimetic
resemblances, 70, 91, 143
Sweet-peas, experiments on, 91
Swynnerton, on contents of
stomachs of birds, 114

Telipna sanguinea, 36
Terias brigitta, 35; *T. hecabe*, 110
Thyridia, 40, **XIV. 3**
Tithorea pseudonyma, **X. 10**
"Transparency group," in S.
America, 39
Trimen, on mimicry in African
butterflies, 13
Tupaia ferruginea, 121

Variation, difficulty of initial,
63

Wade, on relative abundance of
the three forms of *P. polytes* in
Ceylon, 99
Wallace, on mimicry in Oriental
butterflies, 12; on the conditions of mimicry, 50; on the
females of *P. polytes*, 76; on
initial variation, 64
Warning colours, 10, 11
Weismann, 1, 2

Printed in the United States
By Bookmasters